Lone Pine Publishing

Perennials *for* Pennsylvania

Ilene Sternberg
Alison Beck

D0841526

Distributed by Lone Pine Publishing
1808 B Street NW, Suite 140
Auburn, WA, USA 98001
Website: www.lonepinepublishing.com

Library and Archives Canada Cataloguing in Publication

Sternberg, Ilene
    Perennials for Pennsylvania / Ilene Sternberg, Alison Beck.

Includes index.
ISBN-13: 978–1–55105–497–1

    1. Perennials—Pennsylvania. I. Beck, Alison, 1971– II. Title.

SB434.S74 2007   635.9'3209748   C2006–903194–0

All photos by Tim Matheson, Tamara Eder, Laura Peters and Allison Penko, with the following exceptions. All photos used with permission.
Karen Carriere 37b, 119a, 191a&b, 192b; Janet Davis 193; Joan de Grey 111a&b, 164, 185b; Therese D'Monte 88–89; Derek Fell 3, 74, 77b, 96a, 97b, 104, 107, 112a, 139a, 140a&b, 143b, 156b, 162, 173, 203b, 205b, 209a&b, 239a, 240b, 260b, 289b, 295b, 301b, 326a, 327a, 329a; Elliot Engley 44a,b,c&d; Erika Flatt 46a,b&c,72a, 73a, 225a, 226a&c, 227c, 262; Kevin Frey 226b, 227a&b; Anne Gordon 241a, 333b; Richard Hawke-CBG 152a, 153a; Liz Klose 81b, 177c, 234a, 277b; Debra Knapke 101b, 149a&b, 165a&b, 297a, 298c; Janet Loughrey 77b, 85a&b, 97c, 113a&b, 155b, 156a, 157b, 239b, 241b, 253a, 260a, 295a&c, 303b, 334a; Marilynn McAra 138, 142b, 325b, 326c; Steve Nikkila 299a&b; Kim Patrick O'Leary 100, 101a, 175a, 281b, 298a; Photos.com 37a, 59b, 61, 112b, 141a, 143a, 283a, 325a, 327b,c&d; PPA 273b; Robert Ritchie 45a, 55b, 62, 65, 90, 91a&b, 92, 93b, 115b, 214, 249a, 286a, 287a; Leila Sidi 1, 288, 324; Peter Thompstone 95b, 163b, 170a, 206c, 228, 248, 249b, 250b, 256a, 286b, 287b, 330; Mark Turner (PR 0017819) 69, (0101956) 77c, (0301452) 96b, (9801168) 139b, (PR 9804364) 154, (9803579) 157a, (PR 0104046) 172, (PR 0104428) 205a, (PR 9804076) 210a, (PR 0104313) 210b, (0013865) 323a, (2682) 323b, (0201801) 326b, 328, 329b, (0017476) 332/334b, (0102219) 238, (0102222) 240a, (PR 0017819) 260–261; Valleybrook Gardens 148, 192–193, 192a, 246a&b; Don Williamson 141b, 142a, 251a, 319b.

PC: 13

Contents

Acknowledgments

Much gratitude goes to Eve Thyrum, who has tested most of these plants in her beautiful garden and whose expertise is invaluable. Thanks.
—*Ilene Sternberg*

Many thanks to everyone involved in the production of this book, in particular Ilene, and the editorial and design staff at Lone Pine, who did such a wonderful job putting the book together, as always. —*Alison Beck*

The Flowers at a Glance

Pictorial Guide in Alphabetical Order

Ajuga
p. 70

Allium
p. 74

Anemone
p. 76

Artemisia
p. 80

Baptisia
p. 90

Black-Eyed Susan
p. 92

Arum
p. 84

Aster
p. 86

Bleeding Heart
p. 94

Blue Star Flower
p. 98

Boltonia
p. 100

Brunnera
p. 102

Bugbane
p. 104

Campanula
p. 106

Catmint
p. 108

Chrysanthemum
p. 110

Columbine
p. 114

Coral Bells
p. 118

Corydalis
p. 128

Culver's Root
p. 130

Cyclamen
p. 132

Coreopsis
p. 124

Daffodil
p. 134

Daylily
p. 140

Dead Nettle
p. 144

Dwarf Plumbago
p. 148

Epimedium
p. 150

Euphorbia
p. 154

False Solomon's Seal
p. 158

Flowering Fern
p. 160

Foamflower
p. 162

Gas Plant
p. 166

Globe Thistle
p. 168

Goat's Beard
p. 170

Green and Gold
p. 174

Hardy Geranium
p. 176

Hardy Orchid
p. 180

Heliopsis
p. 182

Hellebore
p. 184

Hens and Chicks
p. 188

Heucherella
p. 190

Hibiscus
p. 192

Holly Fern
p. 196

Hosta
p. 198

Indian Pink
p. 202

Iris
p. 204

Jacob's Ladder
p. 212

Joe-Pye Weed
p. 214

Lady's Mantle
p. 216

Jack-in-the-Pulpit
p. 208

Lamb's Ears
p. 220

Ligularia
p. 222

Lily
p. 224

Lily of the Valley
p. 228

Lilyturf
p. 230

Lungwort
p. 232

Lupine
p. 236

Mayapple
p. 238

Maidenhair Fern
p. 242

Meadow Rue
p. 244

Meadowsweet
p. 248

Mondo Grass
p. 252

Monkshood
p. 254

Oriental Poppy
p. 258

Ostrich Fern
p. 262

Pachysandra
p. 264

Painted Fern/Lady Fern
p. 266

Pasqueflower
p. 268

Penstemon
p. 270

Peony
p. 274

Phlox
p. 280

Pinks
p. 284

Primrose
p. 288

Purple Coneflower
p. 292

Sage
p. 296

Sea Holly
p. 300

Sedge
p. 302

Sedum
p. 304

Shasta Daisy
p. 308

Stoke's Aster
p. 316

Solomon's Seal
p. 310

Speedwell
p. 312

Sunflower
p. 318

Toad Lily
p. 320

Trillium
p. 322

Tulip
p. 324

Virginia Bluebells
p. 328

Wild Ginger
p. 330

Wood Fern
p. 332

Yarrow
p. 336

Introduction

Pennsylvania's Horticultural History

Pennsylvania has often been called "The Gateway to America's Gardens." Since colonial times, its significance to horticulture has been renowned. Bartram's Garden in Philadelphia is a National Historic Landmark and our country's oldest surviving botanical garden. Originally, it was the home of America's first native-born botanist, John Bartram (1699-1777). He and his son, William, wandered the previously unexplored eastern wilderness discovering and collecting New World plants. In 1803, at the behest of avid gardener and third U.S. President Thomas Jefferson, Meriwether Lewis and William Clark assembled a team in Philadelphia to embark on an expedition across the uncharted continent. One of their purposes was to collect and identify native plants.

A Philadelphia seed house established by David Landreth in 1784 and continuing for several generations was America's first important business dealing exclusively in seeds. By the time of the Civil War, it was exporting plants around the world. Many other early Philadelphia seed purveyors and nurserymen provided gardeners in the United States and around the world with stock. R-P Nurseries in Kennett Square, first opened in 1866, is the state's oldest continuously operating garden center.

Pennsylvania has more public gardens and arboreta than any other state in the union. It is home to the Pennsylvania Horticultural Society, the first U.S. organization of its kind, and the sponsor of the Philadelphia Flower Show. It is the country's oldest and the world's largest indoor flower show, and is visited by some 300,000 people each year.

Early Quakers built modest homes and surrounded them with outstanding gardens, setting the standard for Pennsylvania's gardening legacy. The Amish, having settled here, have long contributed to the state's horticultural tradition by farming, gardening and establishing excellent plant nurseries.

Families with horticultural interests such as the duPonts, Wistars, Barneses, Pennocks, Copelands, Henrys and others have done much to further the pursuit of gardening and horticultural education, in addition to designating their own estates for public enjoyment.

Perennial plants, naturally, are part of Pennsylvania's horticultural heritage. We are fortunate to be able to see, learn about, buy and grow them here with great ease.

Perennials are plants that take three or more years to complete their life cycle. This is a broad definition that includes trees and shrubs. To narrow the definition in the garden, we refer to herbaceous perennials as perennials. Herbaceous perennials live for three or more years, but they generally die back to the ground at the end of the growing season and start fresh with new shoots each spring. There are some plants that are grouped with perennials that do not die back completely and still others that remain green all winter. Sub-shrubs, like thyme, and evergreen perennials, such as pinks, are examples of other plants grouped with perennials.

Seasonal conditions vary in Pennsylvania, providing unique challenges in every garden. No matter how difficult the site, there are perennials that will flourish and provide the gardener with an almost limitless selection of colors, sizes and forms. This versatility, along with the beauty and permanence of perennials, lies at the root of their continued and growing popularity.

Climate

The temperate climate of Pennsylvania, with its warm summers, cold winters and fairly dependable rainfall, is ideal for growing a wide range of perennials. Added to this is the enthusiasm and creativity of the many people involved in gardening in Pennsylvania. From one end of the state to the other there are individuals, growers, breeders, societies, schools, publications and public gardens all devoted to the advancement and enjoyment of gardening in Pennsylvania. These people are happy to provide information, encouragement and fruitful debate for the gardener; they nurture a knowledge of planting and propagation methods, a precision in identifying specific plants, and a plethora of passionate opinions on what is best for any little patch of ground. Watch for the Home and Garden Show schedules and Spring and Fall Flower Festivals in your local newspaper. For design ideas and inspiration visit one of the many public gardens that abound in Pennsylvania. See p. 67, "Gardens to Visit," to help you plan your garden explorations.

Hardiness Zones Map

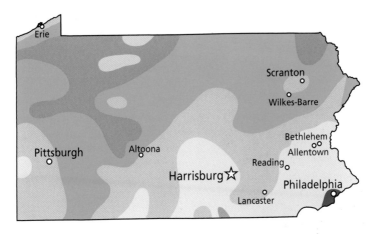

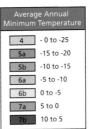

Average Annual Minimum Temperature	
4	- 0 to -25
5a	-15 to -20
5b	-10 to -15
6a	-5 to -10
6b	0 to -5
7a	5 to 0
7b	10 to 5

According to the 1990 USDA Plant Hardiness Map, Pennsylvania falls mostly in zones 5 and 6, though some colder and some warmer areas are also represented. For the last spring frost date, averages across the state fall between April 15 and May 20, with the later dates occurring in rural areas and in the mountains. At the end of the season, the average first frost dates fall between September 30 and October 25.

The degree to which our plants are heat stressed depends on the maximum summer temperatures and nighttime temperatures. Days when the daytime temperature is above 90° F and the nighttime temperatures stay above 70° F not only stress our plants, but us, too! Here in Pennsylvania the number of days and nights when this occurs ranges from about 14 to 45, with some areas experiencing such heat more often than others.

Precipitation averages range from 34–50" annually. Rain in summer is fairly consistent, but periods of drought are possible and can stress plants that require a steady supply of moisture. Mulching can reduce moisture loss and reduce the amount of watering that may be required during dry spells.

Pennsylvania's snowfall has been extremely variable in recent years. Snow provides a blanket for dormant and evergreen plants. Cold temperatures without snow cover are a major cause of winter damage and winter-kill. Periodically, Pennsylvanian's will experience wet winters where it often rains more than it snows. This can create a problem for perennials. One of the leading causes of death of many perennials

Soil

If you get a group of gardeners together the conversation will eventually turn to a discussion of soils and the Herculean efforts that the gardener had to accomplish to create his or her garden. Soil types vary greatly throughout Pennsylvania because of the vast range of regions with the varied influences from glacial deposits in the northwest across the Appalachians to the inland edges of the coastal floodplains in the southeast.

Microclimates

Microclimates abound, giving gardeners almost everywhere the possibility of growing that one perennial that everyone says won't grow here. The challenge of gardening with plants that are borderline hardy is part of the fun of growing perennials, so don't let information about climate zones and perennial hardiness limit your experimentation with plants. The microclimate of your own garden will be influenced by the nearness of buildings, how quickly the soil drains and whether you garden in a low, cold hollow or on top of a windswept knoll or hillside. Unlike trees and shrubs, perennials are relatively inexpensive and easy to share with friends and neighbors, so the more varieties you try, the more likely you'll be to discover what loves to grow in your garden. When it comes to perennials, the best advice is to dig in and "just grow for it."

is "wet feet" in the winter. As you peruse the plant selections, please notice how many of them state that a well-drained site is important for survival. Many plant deaths are attributed to winter cold because they do not reappear in the spring. But often it is winter wet, not winter cold, that kills them.

Another characteristic of Pennsylvania's weather is the frequent freeze/thaw cycles that occur from December through February. It is not uncommon for the mercury to drop into the "teens" in mid-December, and then climb to balmy 40-50° temperatures by Christmas. This frequently occurs in late January and early February. For some plants, such as the subshrubs, short periods of warm temperatures are a signal to start growing. When the temperature drops below freezing, the growth processes that began are stopped, but not without a cost to the plant in wasted food reserves.

Perennial Gardens

A good perennial garden is about more than just flowers and can be interesting throughout the entire year. Consider the foliage of the perennials you want to use. Foliage can be bold or flimsy, coarse or refined; it can be big or small, light or dark; its color can vary from yellow, gray, blue or purple to any multitude of greens; and it can be striped, splashed, edged, dotted or mottled. The texture can be shiny, fuzzy, silky, rough or smooth. The famous white gardens at Sissinghurst, England, were designed not to showcase a haphazard collection of white flowers, but to remove the distraction of color and allow the eye to linger on the foliage, to appreciate its subtle appeal. Flowers come and go, but a garden planned with careful attention to foliage will always be interesting.

Using and Selecting Perennials

Perennials can be used alone in a garden or combined with other plants such as trees, shrubs and annuals. Perennials form a bridge in the garden between the permanent structure provided by trees and shrubs and the temporary color provided by annuals. They often flower for longer and grow to mature size more quickly than shrubs do, and in many cases require less care and are less prone to pests and diseases than annuals.

Many coral bells have colorful foliage.

Perennials can be included in any type, size or style of garden. From the riot of color in a cottage garden or the cool, soothing shades of green in a woodland garden, to a welcoming cluster of pots on a front doorstep, perennials open up a world of design possibilities for even the most inexperienced gardener.

It is very important when planning your garden to decide what you like. If you enjoy the plants that are in your garden, then you are more likely to take proper care of them. Decide on what style of garden you like as well as what plants you like. Think about the gardens you have most admired in your neighborhood, in books or while visiting friends. Use these ideas as starting points for planning your own garden.

Select perennials that flower at different times in order to have some part of your garden flowering all season. In Pennsylvania it is possible to have perennials blooming from mid-February (Lenten rose) to late November (chrysanthemums). (See Quick Reference Chart, p. 340.)

Next, consider the size and shape of the perennials you choose. Pick a variety of forms to make your garden more interesting. The size of your garden influences these decisions, but do not limit a small garden to small perennials or a large garden to large perennials. Use a balanced combination of plant sizes that are in scale with their specific location. (See Quick Reference Chart, p. 340.)

Color...tomes have been written on the subject of color in art and design. We tend to focus on color because it is the first impression that we receive of a garden. Choose a variety of flower colors for your garden, but do not forget that wonderful foliage effects can be achieved by varying leaf color. And, remember, flowers are more ephemeral in their beauty than foliage. Different colors have different effects on our senses. Cool colors like blue, purple and green are soothing and make small spaces seem bigger as they appear to move away from the viewer. Warm colors like red, orange and yellow are more stimulating and appear to fill large spaces as they visually move towards the viewer. (See Quick Reference Chart, p. 340.)

Textures can also create a sense of space. Larger leaves are considered bold or coarse in texture and their visibility from a greater distance makes spaces seem smaller and more shaded. Small leaves, or those that are finely divided, are considered fine in texture and create a sense of greater space and light. Some gardens are designed solely by texture.

Bold-textured Perennials
Allium
Hibiscus
Hosta
Goat's beard
Joe-Pye weed
Ligularia
Lungwort
Mayapple
Peony
Purple coneflower
Sedum 'Autumn Joy'

Fine-textured Perennials
Artemisia
Bleeding heart
Columbine
Coreopsis
Corydalis
Meadow rue
Painted fern

Hosta (above), Japanese painted fern (center)

Decide how much time you will have to devote to your garden. With good planning and advanced preparation, you can enjoy lower-maintenance perennial gardens. Consider using plants that perform well with little maintenance and ones that are generally pest and disease free.

Low-maintenance Perennials
Ajuga
Catmint
Coral bells
Daylily
Dead nettle
Dwarf plumbago
Hardy geranium
Hosta
Pinks
Sedum 'Autumn Joy'
Shasta daisy
Stokes' aster

Lamium (below)

Getting Started

Once you have some ideas about what you want in your garden, consider the growing conditions. Plants grown in ideal conditions, or conditions as close to ideal as you can get them, are healthier and less prone to pest and disease problems than plants growing in stressful conditions. Some plants considered high maintenance become low maintenance when grown in the right conditions.

Avoid trying to make your garden match the growing conditions of the plants you like. Choose plants to match your garden conditions. The levels of light, the type of soil and the amount of exposure in your garden provide guidelines that make plant selection easier. A sketch of your garden, drawn on graph paper, may help you organize the various considerations you want to keep in mind as you plan. Start with the garden as it exists. Knowing your growing conditions can prevent costly mistakes—plan ahead rather than correct later.

Light

Buildings, trees, fences and even the time of day all influence the amount of light that gets into your garden. There are four basic categories of light in the garden: full sun, partial shade, light shade and full shade. Knowing what light is available in your garden helps in deciding where to put each plant.

Full sun locations receive direct sunlight all or most of the day (more than six hours/day). An example is a location along a south-facing wall. Partial shade, or partial sun, locations receive direct sun for part of the day

and shade for the rest (four to six hours/day). An east- or west-facing wall usually gets only partial sun. Light shade locations receive shade most or all of the day, but some sun gets through to ground level. The ground under a small-leaved tree is often lightly shaded. Small dapples of sun are visible on the ground beneath the tree. Areas with full shade receive no direct sunlight. The north side of a house or beneath a densely leaved tree are examples of full shade locations.

It is important to remember that the intensity of the full sun can vary. For example, between buildings in a city, heat can become trapped and magnified, baking all but the most heat-tolerant of plants in a concrete oven. Conversely, the sheltered hollow in the shade that protects your heat-hating plants in the humid summer heat may become a frost trap in winter, never touched by the warmth of the sun and killing tender plants that should otherwise survive.

Daylily (above), Sedum (bottom left), Foamflower (below)

Perennials for Full Sun
Artemisia
Black-eyed Susan
Coreopsis
Daylily
Hibiscus
Oriental poppy
Phlox
Sedum
Sunflower

Perennials for Full Shade
Bleeding heart
Dead nettle
Foamflower
Hosta
Lily-of-the-valley
Lungwort
Mayapple
Monkshood
Primrose
Solomon's seal
Wild ginger

Soil

Plants and the soil they grow in have a unique relationship. Many plant functions go on underground. Soil holds air, water, nutrients, organic matter and a variety of beneficial organisms. Plant roots depend upon these resources while using the soil to hold themselves upright.

Soil is made up of particles of different sizes. Sand particles are the largest. Water drains quickly out of sandy soil and nutrients are quickly washed away. Sand has lots of air space and doesn't compact easily. Clay particles are the smallest and can only be seen through a microscope. Water penetrates clay very slowly and drains very slowly. Clay holds the most nutrients, but there is very little room for air, and clay compacts quite easily. Most soil is made up of a combination of different particle sizes. These soils are called loams.

Perennials for Sandy Soil

Artemesia
Aster
Baptisia
Daylily
Euphorbia
Heliopsis
Sedum
Stokes' aster
Yarrow

Heliopsis

Perennials for Clay Soil

Ajuga
Black-eyed Susan
Boltonia
Hardy geranium
Green and gold
Foamflower
Hosta
Lady's mantle
Lily-of-the-valley
Purple coneflower
Wild ginger
Yarrow

Soil structure is the arrangement of the mineral and organic components into stable aggregates. The aggregation process is aided by roots pushing through the soil and from the substances the roots secrete, by wetting and drying, by freezing and thawing, and especially by soil microorganisms in their search for food.

The organic matter portion of the soil includes living and dead components. The dead components are the residues, metabolites and waste products of plants, animals and microorganisms. The living components are composed of bacteria, fungi, nematodes, protozoa, arthropods and earthworms. These organisms decompose organic compounds, including plant and animal residues and wastes, pesticides and other pollutants. They trap and store nitrogen and other nutrients in their bodies, and they produce hormones that plants use. Their activities enhance soil structure for better air and water movement into the soil and less runoff. They compete with and prey on plant pests, and they provide food for above-ground animals. The living soil organisms, the dead organic matter components, the plants and the above-ground animals make up what is known as the soil food web.

The other aspect of soil to consider is the pH. This is the scale on which the acidity or alkalinity is analyzed. Soils in Pennsylvania vary from acidic to alkaline. You can test your soil if you plan to amend it. Testing kits are available at most garden centers. Soil acidity influences which nutrients are available for plants. Soil can be made more alkaline with the addition of horticultural lime. Soil acidity can be increased with the addition of things like sulphur, peat moss, pine needles or chopped oak leaves. Altering the pH of your soil takes a long time, often many years, and is not easy. If there are only one or two plants you are trying to grow that require a soil that is more, or less, acidic consider growing them in a container or a raised bed where it will be easier to control and amend the pH as needed. Most plants prefer soil pH between 5.5 and 7.5.

Lungwort

Another thing to consider is how quickly the water drains out of your soil. Rocky soil on a hillside will probably drain very quickly and should be reserved for those plants that prefer a very well-drained soil. Low-lying areas tend to retain water longer, and some areas may rarely drain at all. Moist areas can be used for plants that require a consistent water supply, and the areas that stay wet can be used for plants that prefer boggy conditions. Drainage can be improved in very wet areas by adding organic matter to the soil and by building raised beds. Water retention in sandy soil can be improved through the addition of organic matter.

Perennials for Moist Soil
Bleeding heart
Goat's beard
Hibiscus
Hosta
Iris
Jack-in-the-pulpit
Lady's mantle
Ligularia
Lungwort
Meadowsweet
Monkshood
Primrose

False Solomon's seal needs acidic soil.

Perennials for Dry Soil
Artemisia
Baptisia
Catmint
Euphorbia
Lamb's ears
Phlox (moss)
Oriental poppy
Pasqueflower
Pinks
Purple coneflower
Sedum
Yarrow

Lamb's ears (left), Globe thistle (above)

Exposure

Finally, consider the exposure in your garden. Wind, heat, cold and rain are some of the elements your garden is exposed to, and different plants are better adapted than others to the potential damage these forces can cause. Buildings, walls, fences, hills, hedges and trees can all influence your garden's exposure.

Wind in particular can cause extensive damage to your plants. Plants can become dehydrated in windy locations because they may not be able to draw water out of the soil fast enough to replace the water that is lost through the leaves. Tall, stiff-stemmed perennials can be knocked over or broken by strong winds. Some plants that do not require staking in a sheltered location may need to be staked in a more exposed one. Use plants that are recommended for exposed locations or temper the effect of the wind with a hedge or some trees. A solid wall will create turbulence on the leeward side, while a looser structure, like a hedge, breaks up the force of the wind and protects a larger area.

Perennials for Exposed Locations
Anemone
Black-eyed Susan
Columbine
Euphorbia
Globe thistle
Hens and chicks
Phlox (creeping)
Sea holly
Sedum (groundcover species)
Yarrow

Map out your garden's various growing conditions, such as shaded areas, wet areas, windy areas and so on. This guideline will help you recognize where your plants will do best.

Preparing the Garden

Taking the time before you start planting to properly prepare the flowerbeds will save you time later on. Removing all weeds and amending the soil with organic matter prior to planting is the first step in caring for your perennials. Thoroughly digging over a bed and picking out all the weeds by hand is the most thorough technique.

Organic Soil Amendments

All soils, from the heaviest clay to the lightest sand, benefit from the addition of organic matter. Some of the best organic additives are compost, well-rotted manure and composted hemlock bark mulch because they add nutrients as well as improve the soil. These additives improve heavy clay soils by loosening them and allowing air and water to penetrate. Organic matter improves sandy or light soils by increasing the ability of the soils to retain water, which allows plants to absorb nutrients before they are leeched away. Mix organic matter into the soil with a garden fork. Within a few months, earthworms and other decomposer organisms will break down the organic matter. At the same time, their activities will keep the soil from compacting.

In forests, meadows or other natural environments, organic debris such as leaves and various plant bits break down on the soil surface and the nutrients are gradually made available to the plants that are growing there. In the home garden, where pests and diseases may be a problem and where untidy debris isn't practical, a compost pile or bin is useful. Compost is a great regular additive for your perennial garden and good composting methods will help reduce pest and disease problems.

Turning compost into beds (above)

Removing weeds and debris before planting (above)

Wooden compost bins

Composting

Compost can be made in a pile, in a wooden box or in a purchased composter, and the process is not complicated. A pile of kitchen scraps, grass clippings and fall leaves will eventually break down if simply left alone. This is called a "passive" or cool compost pile. It may take one to two seasons for the materials to break down. The process can be sped up if a few simple guidelines are followed to create an "active" or hot compost pile.

Use dry as well as fresh materials with a higher proportion of dry matter than fresh green matter. Dry matter may be chopped straw, shredded leaves or sawdust, whereas green matter may be vegetable scraps, grass clippings or pulled weeds. The green matter breaks down quickly and produces nitrogen, which composting organisms use to break down dry matter. Spread the green materials evenly throughout the pile by layering them between dry materials.

Layers of soil or finished compost will introduce the organisms that are necessary to break down the compost pile properly. If the pile seems very dry you can add a bit of water as you layer. The pile needs to be moist but not soggy.

Turn the pile over or poke holes in it with a pitch fork every week or two. Air must get into the pile and will speed up decomposition. A well-aerated compost pile will generate a high degree of heat. A thermometer

Different styles of plastic compost bins

attached to a long probe, similar to a meat thermometer, will be able to take the temperature near the middle of the pile. Compost can easily reach temperatures of 160° F while decomposing. At this temperature weed seeds have been destroyed and many damaging soil organisms killed. Most beneficial organisms are not killed unless the temperature rises above this temperature. Once your pile reaches 160° F, turn the pile to aerate it and prevent the temperature from getting any higher.

Your compost has reached the end of its cycle when you can no longer recognize the matter that went into it and the temperature no longer rises when you turn the pile. It may take as little as one month to reach this stage and be ready to spread onto your perennial garden. It will have a good mixture of nutrients and be rich in beneficial organisms.

Avoid putting diseased, pest-ridden material into your compost pile. By adding this material, you risk spreading problems throughout your entire garden. If you do put questionable material in the pile, put it as near the center as possible where the temperatures are highest. Do not put diseased or pest-ridden material into a passive compost pile. The internal temperature does not rise high enough to kill pests and diseases.

Green and brown materials for composting

An active (hot) compost process

Finished compost ready to use

Compost worms

Selecting Perennials

Plants can be purchased or started from seed. Purchased plants may begin flowering the same year they are planted, while plants started from seed may take several years to mature. Starting plants from seed is more economical if you want large numbers of plants. (See how to start seeds in the Propagation section, p. 43.)

Plants and seeds are available from many sources. Garden centers, mail order catalogs and even friends and neighbors are excellent sources of perennials. Get your perennials from a reputable source and be sure that the plants are not diseased or pest-ridden.

As well as garden centers, there are a number of garden societies that promote the exchange of seeds, and many public gardens sell seeds of rare plants. Gardening clubs are also a great source of rare and unusual plants.

Purchased perennials come in two main forms. They are sold in pots or they are sold bare-root, usually packed in moist peat moss or sawdust. Potted perennials are growing and have probably been raised in the pot. Bare-root perennials are typically dormant, although some of the previous year's growth may be evident, or there may be new growth starting. Sometimes the piece of root

A rootbound plant (left) and a healthy plant (right)

appears to have no evident growth, past or present. Both potted and bare-root perennials are good purchases, and in each case there are things to look for to make sure that you are getting a plant of the best quality.

Potted plants come in many sizes, and though a larger plant may appear more mature, it may be better to choose a smaller one that will suffer less from the shock of being transplanted. Most perennials grow quickly once they are planted in the garden. Select plants that seem to be a good size for the pot they are in. When tapped lightly out of the pot, the roots should be visible but not winding and twisting around the inside of the pot. The leaves should be a healthy color.

If the leaves appear to be chewed or damaged, check carefully for insects or diseases before you purchase the plant. If you find insects on the plant you may not want to purchase it unless you are willing to cope with the hitchhikers you are taking home. If the plants are diseased, do not purchase them. Deal with any pest problems before you move the plants into the garden, to avoid spreading the pest.

Once you get your plants home, water them if they are dry, and keep them in a lightly shaded location until you plant them. Remove any damaged growth and discard it. Plant your new perennials into the garden as soon as possible. If you have taken advantage of an end-of-the-season sale, you will probably have to deal with rootbound plants. Before planting you will need to

Root mass of a rootbound plant

tease apart the roots if they are winding around the rootball. If there is a solid mat at the bottom of the rootball, remove it. Those roots will not be able to spread out and establish themselves in the soil. New root tips will only be trapped in the existing mass. Gently spread out the roots as you plant. And, last, but not least, water the plant in well. Fall planted perennials may be subject to frost heaving because they were planted later in the season and did not have an opportunity to establish a good root system. When the ground cools—after several frosts and extended cool temperatures—you may want to mulch plants that were set in the garden in the fall to avoid heaving.

Bare-root plants are most commonly sold through mail order, but some are available in garden centers, usually in the spring. Choose roots that are dormant (without top growth) because a plant may take longer to establish itself if it is growing before being placed in the garden. It may have too little energy to recover after trying to grow in the stressful conditions of a plastic bag.

Cut off any damaged parts of the roots with a very sharp knife. Bare-root perennials need to be planted more quickly than potted plants because they will dehydrate quickly out of soil. Soak the roots in luke-warm water for 1–2 hours to rehy-drate them. Do not leave them in water longer as you may promote the development of root or crown rot. Then, either plant them directly in the garden or into pots with good quality potting soil until they can be moved to the garden.

It is often difficult to distinguish the top from the bottom of some bare-root plants. Usually there is a tell-tale dip or stub from which the plant previously grew. If you can't find any distinguishing characteris-tics, lay the root in the ground on its side and the plant will send the roots down and the shoots up.

Perennials lined up for sale

Planting Perennials

Once you have your garden planned, the soil well prepared and the perennials ready, it is time to plant. If your perennials have identification tags, be sure to poke them into the soil next to the newly planted perennials. Next spring, when most of your perennial bed is nothing but a few stubs of green, the tags will help you with identification and remind you that there is indeed a plant in that bare patch of soil.

Potted Perennials

Perennials in pots are convenient because you can space them out across the bed or rearrange them before you start to dig. Once you have the collection organized you can begin planting. Do not unpot the plant until immediately before you transplant to prevent the roots from drying out.

To plant potted perennials, start by digging a hole about the width and depth of the pot. Remove the perennial from the pot. If the pot is small enough, you can hold your hand across the top of the pot, letting your fingers straddle the stem of the plant, and then turn it upside-down. Never pull on the stem or leaves to get a plant out of a pot. It is better to

Support the plant as you remove the pot.

Loosen the rootball a little before planting.

cut a difficult pot off rather than risk damaging the plant. Tease a few roots out of the soil ball to get the plant growing in the right direction. If the roots have become densely wound around the inside of the pot, you should cut into the root mass with a sharp knife to encourage new growth into the surrounding soil. The process of cutting into the bottom half of the root ball and spreading the two halves of the mass outward like a pair of wings is called butterflying the roots and is a very effective way to promote fast growth of pot-bound perennials when transplanting. Place the plant into the prepared hole. It should be buried to the same level that it was at in the pot, or a little higher, to allow for the soil to settle. If the plant is too low in the ground, it may rot when rain collects around the crown. Fill the soil in around the roots and firm it down. Water the plant well as soon as you have planted it, and regularly until it has established itself.

Bare-root Perennials

During planting, bare-root perennials should not be spaced out across the bed unless you previously planted them in temporary pots. Roots dry out very quickly if you leave them lying about waiting to be planted. If you want to visualize your spacing, you can poke sticks into the ground or put rocks down to represent the locations of your perennials.

If you have been keeping your bare-root perennials in potting soil, you may find that the roots have not grown enough to knit the soil together and that all the soil falls away from the root when you remove it from the pot. Don't be concerned. Simply follow the regular root-planting instructions. If the soil does hold together, plant the root the way you would a potted perennial.

Root Types

The type of hole you need to dig will depend on the type of roots the perennial has. Plants with fibrous roots will need a mound of soil in the center of the planting hole over which the roots can be spread out evenly. The hole should be dug as deep as the longest roots. Mound the soil into the center of the hole up to ground level. Spread the roots out around the mound and cover them with loosened soil. If you are adding just one or two plants and do not want to prepare an entire bed, dig a hole twice as wide and deep as the root ball and amend the soil with composted manure mixed with peat moss. Add a slow release organic fertilizer to the backfill of soil that you spread around the plant. Composted chicken or barnyard manure can also be used to improve planting areas. Granulated, composted chicken manure is a wonderful product that is available from suppliers of organic gardening products. Horse manure is also an excellent additive and is usually available from stables, often in a seemingly endless supply. If you have access to fresh manure, compost it first or use it sparingly because roots that come in contact with fresh manure will suffer fertilizer burn. Consider incorporating it into

Fill in soil around roots and firm into place (above). Water in new plantings immediately (below).

beds that you prepare at least two weeks—or better yet—the season before planting. Avoid using it in vegetable beds at all to prevent the potential spread of fecal-borne diseases.

Plants with a tap root need a hole that is narrow and about as deep as the root is long. The job is easily done with the help of a trowel: open up a suitable hole, tuck the root into it and fill it in again with the soil around it. If you can't tell which end is up, plant the root on its side.

Some plants have roots that appear to be tap roots, but the plant seems to be growing off the side of the root, rather than upwards from one end. These roots are called rhizomes. Iris roots are rhizomes. Rhizomes should be planted horizontally in a shallow hole and covered with soil.

In most cases, you should try to get the crown at or just above soil level and loosen the surrounding soil in the planting hole. Keep the roots thoroughly watered until the plants are well established.

Whether the plants are potted or bare-root, it is good to leave them alone and let them recover from the stress of planting. In the first month, you will need only to water the plant regularly, weed it and watch for pests. A mulch spread on the bed around your plants will keep in moisture and control weeds.

If you have prepared your beds properly, you probably won't have to fertilize in the first year. If you do want to fertilize, wait until your new plants have started healthy new growth, and apply only a weak fertilizer to avoid damaging the new root growth.

Planters

Perennials can also be grown in planters for portable displays that can be moved about the garden. They can be used on patios or decks, in gardens with very poor soil or in yards where kids and dogs might destroy a traditional perennial bed. Many perennials such as hosta and daylily can grow in the same container without any fresh potting soil for five or six years. Be sure to fertilize and water perennials in planters more often than those growing in the ground. Dig your finger deep

into the soil around the perennial to make sure it needs water. Too much water in the planter can encourage root rot.

Always use a good quality potting mix or a soil mix intended for containers in your planters. Garden soil quickly looses its structure and becomes a solid lump in a container. This prevents air, water and roots from penetrating into the soil. Plants will never thrive in smaller containers if planted entirely in soil from the garden. In larger containers (e.g., top diameter of 36"), using some garden soil will help stabilize the growing media. At the very least, mix half garden soil with half peat moss, and mix the two together well.

When designing a planter garden, you can either keep one type of perennial in each planter and display many planters together, or mix different perennials in large planters along with annuals and bulbs. The latter choice results in a dynamic bouquet of flowers and foliage. Keep tall, upright perennials, such as yarrow, in the center of the planter; rounded or bushy types like heuchera around the sides; and low-growing or draping perennials, such as the species candytuft, along the edge of the planter. Perennials that have long bloom times or attractive foliage are good for planters.

Choose hardy perennials that are tolerant of difficult conditions. Planters are exposed to extremes of our variable weather: baking hot in summer and freezing cold in winter. They dry out quickly in hot weather and become waterlogged after a couple of rainy days. Not all perennials

Lady's mantle

are tough enough to survive in these extreme conditions. Some of the more invasive perennials are a good choice, because their spread is controlled, but at the same time they are very tough to kill.

Perennials in planters are more susceptible to winter damage because the exposed sides of the container allow for greater fluctuations in temperature and provide very little insulation for roots. The container itself may even crack when exposed to a deep freeze. Don't despair, there are plenty of things you can do to get planters through a tough winter in great shape. The simplest thing you can do is move the planter to a more sheltered spot. Most perennials do require some cold in the winter in order to flower the next year, so find a spot that is still cold, but provides some shelter. An unheated garage or enclosed porch are good places; even your garden shed will offer your plants more protection than they would get sitting in the great outdoors, exposed to the elements on all sides.

If you haven't the space or access to these places, consider your basement window wells. These are sheltered, below ground and have a nearby source of some heat from the window. Wait until your pots have frozen. Layer straw at the bottom of the well, sit your pots on the straw, then cover them with more straw. Waiting until the pots freeze prevents rot and can keep the mice from easily eating the succulent treats you are conveniently leaving out for them. Mice will be the biggest problem. They find the straw makes a comfortable home and the perennial roots to be a tasty treat. Frozen soil is harder to dig into. If mice are a problem you may want to consider sprinkling some commercial mouse bait around the pots. If you have pets or small children around and don't want to use poisons, then a layer of Styrofoam insulation under and on top of the pots will also provide protection, without being quite so appealing to small rodents.

The pots themselves can be winter-proofed before you plant your perennials. Using Styrofoam insulation or "packing peanuts," put a layer at the bottom of the pot and around the inside edge of the planter before you add your soil and plants. Make sure excess water can still drain freely from the container. There are also commercial planter insulating materials available at garden centers. These materials are particularly useful for high-rise dwellers with balcony or rooftop gardens. This insulation will also protect the roots from heat build up in the summer.

Finally, planters can be buried in the garden for the winter. Find an open space in a flower bed, and dig a hole deep enough to allow you to sink the planter up to its rim. This can be messy, particularly in the spring when you dig it up. It also requires that enough empty space be available in the garden to fit your planter or planters. Large planters may require an extensive excavation or even the use of a backhoe in order to dig deeply enough to fit the entire pot, making this option very impractical for all but smaller containers.

Perennials for Planters

Artemisia
Catmint
Coral bells
Daylily
Dead nettle
Hardy geranium
Goat's beard
Hosta
Iris
Lady's mantle
Pinks
Sage
Sedum
Speedwell
Yarrow

Artemisia

Caring for Perennials

Many perennials require little care, but all will benefit from a few maintenance basics. Weeding, pruning, deadheading and staking are just a few of the chores that, when done on a regular basis, keep major work to a minimum.

Weeding

Controlling weeds is one of the most important things you will have to do in your garden. Weeds compete with your perennials for light, nutrients and space. Weeds can also harbor pests and diseases. Try to prevent weeds from germinating. Many weed seeds, especially annual weeds, need light to germinate. Therefore, block their access to light. If they do germinate, pull them out while they are still small and before they have a chance to flower, set seed and start a whole new generation of problems.

Weeds can be pulled out by hand or with a hoe. Quickly scuffing across the soil surface with the hoe will pull out small weeds and sever larger ones from their roots. A layer of mulch is an excellent way to suppress weeds. Also, as your garden matures, the perennials themselves will suppress weeds by blocking the light needed for germination.

Mulching

Mulches are an important gardening tool. They prevent weed seeds from germinating by blocking out the light. Soil temperatures remain more consistent and more moisture is retained under a layer of mulch. Mulch also prevents soil erosion during heavy rain or strong winds.

Mulch helps keep soil moist and weeds at bay.

material like chopped straw, pine needles or shredded leaves. Keep in mind that as the ground freezes, so too may your pile of potential mulch. One solution is to cover most of the bed with mulch, leaving only the plants exposed, before the ground freezes. Put extra mulch, needed to cover the plants, in a large plastic bag or your wheelbarrow and put it somewhere that will take longer to freeze—perhaps in your garage or the garden shed. Once the ground is completely frozen you will have a supply of mulch that you can use to cover the plants.

In late winter or early spring, once the weather starts to warm up, pull the mulch layer off the plants and see if they have started growing. If they have, you can pull the mulch back, but keep it nearby in case you need to put it back on to protect the tender new growth from a late frost. Once your plants are well on their way and you are no longer worried about frost, you can remove the protective mulch completely from the plant crowns, spread some around in the existing bed for your spring and summer mulch and compost the rest. Or you can use it elsewhere to start a new garden bed.

because they improve the soil and add nutrients as they break down.

In spring, spread a couple inches of mulch over your perennial beds around your plants. Keep the area immediately around the crown or stem of your plants clear (2–4" clearance). Mulch that is too close to your plants can trap moisture and prevent good air circulation, encouraging disease. If the layer of mulch disappears into the soil over summer, you should replenish it.

A fresh layer of mulch, up to 3–4" thick, can be laid once the ground freezes in fall, to protect the plants over winter. This is particularly important if you can't depend on a steady layer of snow to cover your garden in the winter, as is the case in parts of Pennsylvania. You can cover the plants with dry

Pruning

Many perennials will benefit from a bit of grooming. Resilient health, plentiful blooming and more compact growth are the signs of a well-groomed garden. Pinching, thinning and disbudding plants before they flower will enhance the beauty of a perennial garden. The methods for pruning are simple, but some experimentation is required in order to get the right effect in your own garden.

Thin clump-forming perennials like black-eyed Susan, coneflower or bergamot early in the year when shoots have just emerged. These plants develop a dense clump of stems that allows very little air or light into the center of the plant. Remove half of the shoots when they first emerge. This removal will increase air circulation and prevent diseases such as powdery mildew. The increased light encourages more compact growth and more flowers. Throughout the growing season, thin any growth that is weak, diseased or growing in the wrong direction.

Trimming or pinching perennials is a simple procedure, but timing it correctly and achieving just the right look can be tricky. Early in the year, before the flower buds appear, trim the plant to encourage new side shoots. Remove the tip and some stems of the plant just above a leaf or pair of leaves. This can be done stem by stem, but if you have a lot of plants you can trim off the tops with your hedge shears to one-third of the height you expect the plants to reach. The growth that begins to emerge can be pinched again. Beautiful layered effects can be achieved by staggering the trimming times by a week or two.

Perennials to Pinch Early in the Season

Artemisia
Aster
Boltonia
Black-eyed Susan
Catmint

Chrysanthemum (above), Hibiscus (below)

Chrysanthemum
Hibiscus
Purple coneflower
Sedum 'Autumn Joy'
Shasta daisy

Give plants enough time to set buds and flower. Continual pinching will encourage very dense growth but also delay flowering. Most spring-flowering plants cannot be pinched back or they will not flower. Early summer or mid-summer bloomers should be pinched only once, as early in the season as possible. Late summer and fall bloomers can be pinched several times but should be left alone past June. Don't pinch the plant if flower buds have formed—it may not have enough energy or time left in the year to develop a new set of buds. Experimentation and keeping detailed notes will improve your pinching skills.

Disbudding is the final grooming stage before a plant blooms. This is the removal of some flower buds to encourage the remaining ones to produce larger flowers. This technique is popular with peony, dahlia and rose growers.

Deadheading—the removal of flowers once they are finished blooming—serves several purposes. It keeps plants looking tidy, prevents the plant from spreading seeds (and therefore seedlings) throughout the garden, often prolongs blooming and helps prevent pest and disease problems.

Flowers can be pinched off, or deadheaded, by hand or snipped off with hand pruners. Bushy plants that have many tiny flowers, particularly ones that have a short bloom period like basket-of-gold, can be more aggressively pruned back with garden shears once they are done flowering. In some cases, e.g., moss phlox, shearing will promote new growth and possibly blooms later in the season.

Coreopsis

Deadheading often promotes more blooms.

Perennials to Shear Back After Blooming

Campanula
Coreopsis
Dead nettle
Hardy geranium
Phlox (creeping)
Yarrow

Perennials with Interesting Seedheads

Baptisia
Columbine
False Solomon's seal
Gas plant
Goat's beard
Meadowsweet

Deadheading is not necessary for every plant. There are some plants with seedheads that are left in place to provide interest in the garden over winter. Other plants are short-lived and by leaving some of the seedheads in place you are encouraging future generations to replace the old plants. Hollyhock is one example of a short-lived perennial that reseeds. In some cases the self-sown seedlings do not possess the attractive features of the parent plant. Deadheading may be required in these cases.

Lupine (above), Pasqueflower (below)

False Solomon's Seal

Oriental poppy
Pasqueflower
Purple coneflower
Sea holly
Sedum 'Autumn Joy'

Ajuga (above), Corydalis (below)

Spiral stakes help support tall, single stems.

Perennials that Self-seed

Ajuga (variable seedlings)
Bleeding heart (variable seedlings)
Corydalis
Lady's mantle
Lupine
Pinks

Staking

Staking, the use of poles, branches or wires to hold plants erect, can often be avoided by astute thinning and pinching, but there are always a few plants that will need a bit of support to look their best in your garden. There are three basic types of stakes to go with the different growth habits that need support. Plants that develop tall spikes, such as hollyhock, delphinium and sometimes foxglove, require each spike to be staked individually. A strong, narrow pole, such as a bamboo stick can be pushed into the ground early in the year and the spike tied to the stake as it grows. A forked branch can also be used to support single-stem plants.

Many plants, such as peony, get a bit top-heavy as they grow and tend to flop over once they reach a certain height. A wire hoop, sometimes called a peony ring, is the most unobtrusive way to hold up such a plant. When the plant is young, the legs of the peony ring are pushed into the ground around it and as the plant grows up, it is supported by the wire ring. At the same time the bushy growth hides the ring. Wire tomato cages can also be used.

Other plants, like coreopsis, form a floppy tangle of stems. These plants can be given a bit of support with twiggy branches that are inserted into the ground around the young plants that then grow up into the twigs.

Some people consider stakes to be unsightly no matter how hidden they seem to be. There are a couple of things you can do to reduce the need for staking. Don't assume a plant will do better in a richer soil than is recommended. Very rich soil causes many plants to produce weak, leggy growth that is prone to lodging (falling over). Also, a plant that likes full sun will be stretched out and leggy if grown in the shade. Plants can give each other some support in the border. Mix in plants

The heavy blooms of peony often need staking.

that have a more stable structure between the plants that need support. A plant may still fall over slightly, but only as far as its neighbor will allow. Many plants are available in compact varieties that don't require staking. But remember, not every plant needs to be ramrod straight. Graceful arching can add another element of interest to the garden.

Watering

Watering is another basic of perennial care. Many perennials need little supplemental watering if they have been planted in their preferred conditions and are given a moisture-retaining mulch. The rule of watering is to water thoroughly and infrequently. If you water frequently and lightly, you will be wetting only the very top layer of soil. Plant roots will stay near the surface instead of growing down. This is causes problems when a dry period hits. Roots that are close to the soil surface cannot take advantage of moist soil that is deeper. So, when you do water, make sure the water penetrates several inches into the soil. One inch of water will infiltrate approximately four inches of soil.

Fertilizing

If you prepare your beds well and add new compost to them each spring, you should not need to add extra fertilizer. Many perennials do not need extra fertilizers. In fact, many of them thrive in poor to average soil. If overfed, many perennials overgrow and become maintenance headaches. If you have a limited amount of compost, you can mix a slow-release fertilizer into the soil around your perennials in the spring. Some plants, e.g., delphinium and hibiscus, are heavy feeders that need additional supplements throughout the growing season.

There are many organic and chemical fertilizers available at garden centers. Be sure to use the recommended quantity because too much fertilizer will do more harm than good. Roots can be burned by fertilizer that is applied in high concentrations. Problems are more likely to be caused by chemical fertilizers because they are more concentrated than organic fertilizers. Most fertilizer instructions recommend a higher rate than is necessary for good plant growth. For perennials it is important to support good root development in the first year or two of growth. Phosphorus is the fertilizer that promotes root growth. So look for fertilizers that are higher in phosphorus. The typical fertilizer formula is N : P : K (Nitrogen : Phosphorus : Potassium). In the years after your plants establish themselves, it is important to provide nitrogen for leaf development and potassium for flower and seed development.

A hose-end watering wand

Propagating Perennials

Learning to propagate your own perennials is an interesting and challenging aspect of gardening that can save you money, but that also takes time and space. Seeds, cuttings and divisions are the three methods of increasing your perennial population. There are benefits and problems associated with each method.

Seeds

Starting perennials from seed is a great way to propagate a large number of plants at a relatively low cost. Seeds can be purchased or collected from your own or a friend's perennial garden. There are some limitations to propagating from seed. Some cultivars and varieties don't pass on their desirable traits to their offspring. Other perennials take a very long time to germinate, if they germinate at all, and an even longer time to grow to flowering size. However, there are many perennials that grow easily from seed and flower within a year or two of being transplanted into the garden. There are

challenges and limitations to starting perennials from seed but the pleasure you will receive when the plants you raised from tiny seedlings finally begin to flower make it worth the work.

Specific propagation information is given for each plant, but there are a few basic rules for starting all seeds. Some seeds can be started directly in the garden but it is easier to control temperature and moisture levels and to provide a sterile environment if you start the seeds indoors. Seeds can be started in pots or, if you need a lot of plants, flats. Use a sterile soil mix intended for starting seeds. The soil will generally

Filling cell packs

Using folded paper to plant small seeds

A spray bottle provides a gentle spray (left).
A prepared seed tray with cover (right)

All seedlings are susceptible to a problem called damping off, which is caused by soil-borne, fungal organisms. An afflicted seedling looks as though someone has pinched the stem at soil level, causing the plant to topple over. The pinched area blackens and the seedling dies. Sterile soil mix, air circulation and evenly moist soil will help prevent this problem.

Fill your pot or seed tray with the soil mix and firm it down slightly—not too firmly or the soil will not drain. Wet the soil before planting your seeds. They may wash into clumps if the soil is watered after the seeds are planted. Large seeds can be placed individually and spaced out in pots or trays. If you have divided inserts for your trays, you can plant one or two seeds per section. Small seeds may have to be sprinkled in a bit more randomly. Fold a sheet of paper in half and place the small seeds in the crease. Gently tapping the underside of the fold will bounce or roll the seeds off the paper in a more controlled manner. Some seeds are so tiny that they look like dust. These seeds can be mixed with a small quantity of very fine sand and spread on the soil surface. These tiny seeds may not need to be covered with any more soil. The medium-sized seeds can be lightly covered, and the larger seeds can be pressed into the soil and then lightly covered. Do not cover seeds that need to be exposed to light in order to germinate. Water the seeds using a very fine spray if the soil starts to dry out. A hand-held spray bottle will moisten the soil without disturbing the seeds.

need to be kept moist but not soggy. Most seeds germinate in moderately warm temperatures of about 57°–70° F.

There are many seed-starting supplies available at garden centers. Some supplies are useful, but many are not necessary. Seed-tray dividers are useful. These dividers, often called plug trays, are made of plastic and prevent the roots from tangling with the roots of the other plants and from being disturbed when seedlings are transplanted. Heating coils or pads can be useful. Placed under the pots or flats, they keep the soil at a constant temperature.

Plant only one type of seed in the pot or flat. Each species has a different rate of germination and the germinated seedlings will require different conditions than the seeds that have yet to germinate. To keep the environment moist, you can place pots inside clear plastic bags. Change the bag or turn it inside out once the condensation starts to build up and drip. Plastic bags can be held up with stakes or wires poked in around the edges of the pot. Many seed trays come with clear, plastic covers that can be placed over the flats to keep the moisture in. Plastic can be removed once the seeds have germinated.

Seeds generally do not require a lot of light in order to germinate, so pots or trays can be kept in a warm, out-of-the-way place. Once the seeds have germinated, they can be placed in a bright location but out of direct sun. Plants should be transplanted to individual pots once they have three or four true leaves. True leaves are the ones that look like the mature leaves. (The first one or two leaves are cotyledons or "seed leaves," which were part of the seed.) Plants in plug trays can be left until neighboring leaves start to touch each other. At this point the plants will be competing for light and should be transplanted to individual pots.

Young seedlings do not need to be fertilized. Fertilizer will cause seedlings to produce soft, spindly growth that is susceptible to attack by insects and diseases. The seed itself provides all the nutrition the seedling will need. A fertilizer, diluted to ¼ or ½ strength, can be used once seedlings have four or five true leaves.

Seeds have protection devices that prevent them from germinating when conditions are not favorable or from all germinating at once. In the wild, staggered germination periods improve the chances of survival. Many seeds will easily grow as soon as they are planted, but others need to have their defenses lowered before they will germinate. Some seeds produce thick seed coats while others produce poisonous chemicals in the seed coats to deter insects.

Perennials to Start from Seed

Boltonia
Columbine
Corydalis
Lady's mantle
Pinks

Columbine (above), Pinks (below)

Soaking seeds in water (above)

Scratching seeds with sandpaper (center)

Seeds can be tricked into thinking the conditions are right for sprouting. Some thick-coated seeds can be soaked for a day or two in a glass of water to promote germination. This mimics the end of the dry season and the beginning of the rainy season, which is when the plant would germinate in its natural environment. The water softens the seed coat and in some cases washes away the chemicals that have been preventing germination. Hibiscus is an example of a plant with seeds that need to be soaked before germinating.

Other thick-coated seeds need to have their seed coats scratched to allow moisture to penetrate the seed coat and prompt germination. In nature, birds scratch the seeds with gravel in their craws and acid in their stomachs. To mimic this effect, nick the seeds with a knife or gently rub them between two sheets of sand paper. Leave the seeds in a dry place for a day or so after scratching them before planting them. This gives the seeds a chance to get ready for germination before they are exposed to water. Lupine and anemone have seeds that need their thick coats scratched.

Plants from northern climates often have seeds that wait until spring before they germinate. These seeds must be given a cold treatment, which mimics winter, before they will germinate. One method of cold treatment is to plant the seeds in a pot or tray and place them in the refrigerator for up to two

Mixing seeds with moist peat moss for cold treatment

Shade cloth covering a cold frame

months. Check the container regularly and don't allow these to dry out. This method is fairly simple, but not very practical if your refrigerator is as crowded as mine. Yarrow, bergenia and primrose have seeds that respond to cold treatment.

A less space-consuming method is to mix the seeds with some moistened sand, peat or sphagnum moss. Place the mix in a sealable sandwich bag and pop it in the refrigerator for up to two months, again being sure the sand or moss doesn't dry out. The seeds can then be planted in the pot or tray. Spread the seeds and the moist sand or moss onto the prepared surface and press it down gently.

A cold frame is a wonderful tool for the gardener. It can be used to protect tender plants over the winter, to start vegetable seeds early in the spring, to harden plants off before moving them to the garden, to protect fall-germinating seedlings and young cuttings or divisions and to start seeds that need a cold treatment. This mini-greenhouse structure is built so that ground level on the inside of the cold frame is lower than on the outside. The angled, hinged lid is fitted with glass. The soil around the outside of the cold frame insulates the plants inside. The lid lets light in and collects some heat during the day and prevents rain from damaging tender plants. If the interior gets too hot, the lid can be raised for ventilation. A hot frame is insulated and has heating coils in the floor to prevent the soil from freezing or to maintain a constant soil temperature for germinating seeds, or rooting cuttings.

Cuttings

Cuttings are an excellent way to propagate varieties and cultivars that you really like but that don't come true from seed or that don't produce seed at all. Each cutting will grow into a reproduction of the parent plant. Cuttings are taken from the stems of some perennials and the roots of others.

Stem cuttings are generally taken in the spring and early summer. During this time, plants go through a flush of fresh, new growth, either before or after flowering. Avoid taking cuttings from plants that are in flower. Plants that are in flower, or are about to flower, are busy trying to reproduce; plants that are busy growing, by contrast, are already full of the right hormones to promote quick root growth. If you do take cuttings from plants that are flowering, be sure to remove the flowers and the buds to divert the plant's energy back into growing.

Large numbers of cuttings don't often result in as many plants. Cuttings need to be kept in a warm, humid place to root, which makes them very prone to fungal diseases. Providing proper sanitation and encouraging quick rooting will increase the survival rate of your cuttings.

Speedwell

Perennials to Propagate from Stem Cuttings
Artemisia
Aster
Boltonia
Bleeding heart
Campanula
Catmint
Coreopsis
Euphorbia
Pinks
Sedum 'Autumn Joy'
Speedwell
Yarrow

Debate exists over what the size of cuttings should be. Some gardeners claim that smaller cuttings are more likely to root and root more quickly. Other gardeners claim that larger cuttings develop more roots and become established more quickly once planted in the garden. You may wish to try different sizes to see what works best for you. A small cutting is 1–2" long and a large cutting is 4–6" long.

The size of cuttings can be determined by the number of leaf nodes on the cutting. You will want at least three or four nodes on a cutting. The node is where the leaf joins the stem, and it is from here that the new roots will grow. The base of the cutting will be just below a node. Strip the leaves gently from the first and second nodes and plant them below the soil. The new plants will grow from the nodes above the soil. The leaves can be left in place on the cutting above ground. If there is a lot of space between nodes, your cutting will be longer than recommended. Some plants have almost no space at all between nodes. Cut these plants to the recommended length and gently remove the leaves from the lower half of the cutting. Plants with several nodes close together often root quickly and abundantly.

Remove the lower leaves.

Always use a sharp, sterile knife to make the cuttings. Cuts should be made straight across the stem. Once you have stripped the leaves, you can dip the end of the cutting into a rooting-hormone powder intended for softwood cuttings. Sprinkle the powder onto a piece of paper and dip the cuttings into it. Discard any extra powder left on the paper to prevent the spread of disease. Tap or blow the extra powder off the cutting. Cuttings caked with rooting hormone do not root any faster than those that are lightly dusted, and they are more likely to rot rather than root. Your cuttings are now prepared for planting.

Dip the cut end in rooting hormone.

Firm the cutting into the soil.

The sooner you plant your cuttings the better. The less water the cuttings lose, the less likely they are to wilt and the more quickly they will root. Cuttings can be planted in a similar manner to seeds. Use sterile soil mix, intended for seeds or cuttings, in pots or trays that can be covered with plastic to keep in the humidity. Other mixes to root the cuttings in are sterilized sand, perlite, vermiculite or a combination of the three. Firm the soil down and moisten it before you start planting. Poke a hole in the surface of the soil with a pencil or similar object, tuck the cutting in and gently firm the soil around it. Make sure the lowest leaves do not touch the soil and that

Newly planted cuttings

Healthy roots on cuttings

the cuttings are spaced far enough apart that adjoining leaves do not touch each other. Pots can be placed inside plastic bags. Push stakes or wires into the soil around the edge of the pot so that the plastic will be held off the leaves. The rigid plastic lids that are available for trays may not be high enough to fit over the cuttings, in which case you will have to use stakes and a plastic bag to cover the tray.

Keep the cuttings in a warm place, about 65°–70° F, in bright, indirect light. A couple of holes poked in the bag will allow for some ventilation. Turn the bag inside out when condensation becomes heavy. Keep the soil moist. A hand-held mister will gently moisten the soil without disturbing the cuttings.

Most cuttings will require from one to four weeks to root. After two weeks, give the cutting a gentle tug. You will feel resistance if roots have formed. If the cutting feels as though it can pull out of the soil, then gently push it back down and leave it for longer. New growth is also a good sign that your cutting has rooted. Some gardeners simply leave the cuttings alone until they can see roots through the holes in the bottoms of the pots. Uncover the cuttings once they have developed roots.

Apply a foliar feed when the cuttings are showing new leaf growth. Plants quickly absorb nutrients through the leaves; therefore, you can avoid stressing the newly formed roots. Your local garden center should have foliar feeds and information about applying them. Your hand-held mister can apply foliar feeds.

Once your cuttings are rooted and have had a bit of a chance to establish themselves, they can be potted up individually. If you rooted several cuttings in one pot or tray, you may find that the roots have tangled together. If gentle pulling doesn't separate them, take the entire clump that is tangled together and try rinsing some of the soil away. This should free the roots enough for you to separate the plants.

Pot the young plants in a sterile potting soil. They can be moved into a sheltered area of the garden or a cold frame and grown in pots until they are large enough to plant in the garden. The plants may need some protection over the first winter. Keep them in the cold frame if they are still in pots. Give them an extra layer of mulch if they have been planted out.

Basal cuttings involve removing the new growth from the main clump and rooting it in the same manner as stem cuttings. Many plants send up new shoots or plantlets around their bases. Often, the plantlets will already have a few roots growing. The young plants develop quickly and may even grow to flowering size the first summer. You may have to cut back some of the top growth of the shoot because the tiny developing roots won't be able to support a lot of top growth. Treat these cuttings in the same way you would a stem cutting. Use a sterile knife to cut out the shoot. Sterile soil mix and humid conditions are preferred. Pot plants individually or place them in soft soil in the garden until new growth appears and roots have developed.

Campanula (above), Garden phlox (below)

Perennials to Start from Tip or Basal Cuttings

Ajuga
Campanula
Catmint
Dead nettle
Euphorbia
Hens and chicks
Hardy geranium
Phlox
Sedum

Root cuttings can also be taken from some plants. Dandelions are well known for this trait: even the smallest piece of root left in the ground can sprout a new plant, foiling every attempt to eradicate them from lawns and flower beds. But there are perennials that have this ability as well. The main difference between starting root cuttings and stem cuttings is that the root cuttings must be kept fairly dry because they can rot very easily.

Cuttings can be taken from the fleshy roots of certain perennials that do not propagate well from stem cuttings. These cuttings should be taken in early or mid-spring when the ground is just starting to warm up and the roots are just about to break dormancy. At this time, the roots of the perennials are full of nutrients, which the plants stored the previous summer and fall, and hormones are initiating growth. You may have to wet the soil around the plant so that you can loosen it enough to get to the roots.

Keep the roots slightly moist, but not wet, while you are rooting them and keep track of which end is up. Roots must be planted in a vertical, not horizontal, position on the soil, and roots need to be kept in the orientation they held while previously attached to the parent plant. There are different tricks people use to recognize the top from the bottom of the roots. One method is to cut straight across the tops and diagonally across the bottoms.

You do not want very young or very old roots. Very young roots are usually white and quite soft; very old roots are tough and woody. The roots you should use will be tan in color and still fleshy. To prepare your root, cut out the section you will be using with a sterile knife. Cut the root into pieces that are one to two inches long. Remove any side roots before planting the sections in pots or planting trays. You can use the same type of soil mix the seeds and stem cuttings were started in. Poke the pieces vertically into the soil and leave a tiny bit of the end poking up out of the soil. Remember to keep the pieces the right way up.

Keep the pots or trays in a warm place out of direct sunlight. Avoid overwatering them. They will send up new shoots once they have rooted and can be planted in the same manner as the stem cuttings (see p. 48).

Rhizomes are the easiest root cuttings with which to propagate plants. Rhizomes are thick, fleshy roots that grow horizontally along the ground, or just under the soil. Periodically, they send up new shoots from along the length of the rhizome. In this way the plant spreads. It is easy to take advantage of this feature. Take rhizome cuttings when the plant is growing vigorously (usually in the late spring or early summer).

Dig up a section of rhizome. If you look closely at it you will see that it appears to be growing in sections. The places where these sections join are called nodes. It is from these nodes that feeder roots (smaller stringy roots) extend downwards and new plants sprout upwards. You may even see that small plants are already sprouting. The rhizome should be cut into pieces. Each piece should have at least one of these nodes in it.

Lily-of-the-valley

Fill a pot or planting tray to about 1" from the top of the container with perlite, vermiculite or seeding soil. Moisten the soil and let the excess water drain away. Lay the rhizome pieces flat on the top of the mix and almost cover them with more of the soil mix. If you leave a small bit of the top exposed to the light, it will encourage the shoots to sprout. The soil does not have to be kept wet, but you should moisten it when it dries out to avoid having your rhizome rot. Once your cuttings have established themselves, they can be potted individually and grow in the same manner as the stem cuttings (see p. 48).

Perennials to Propagate from Rhizomes

Campanula
Hardy geranium
Iris
Lily-of-the-valley
Wild ginger

Division

Division is quite possibly the easiest way to propagate perennials. As most perennials grow, they form larger and larger clumps. Dividing this clump once it gets big will rejuvenate the plant, keep its size in check and provide you with more plants. If a plant you really want is expensive, consider buying only one because within a few years you may have more than you can handle.

How often a perennial needs dividing or can be divided will vary. Some perennials, like astilbe, need dividing almost every year to keep them vigorous, while others, like peony, should

never be divided, because they dislike having their roots disturbed. Each perennial entry in the book gives recommendations for division. There

Digging up perennial clump for division (above, center)

Clump of stems, roots & crowns (below)

Pulling a clump apart

Cutting apart and dividing tuberous perennials

pried apart with a pair of garden forks inserted back to back into the clump. Plants with thicker tuberous or rhizomatous roots can be cut into sections with a sharp, sterile knife. In all cases, cut away any old sections that have died out and replant only the newer, more vigorous sections.

Once your original clump is divided into sections, replant one or two of them into the original location. Take this opportunity to work organic matter into the soil where the perennial was growing before replanting it. The other sections can be moved to new spots in the garden or potted up and given away as gifts to gardening friends and neighbors. Get the sections back into the ground as quickly as possible to prevent the exposed roots from drying out. Plan where you are going to plant your divisions and have the spots prepared before you start digging up. Plant your perennial divisions in pots if you aren't sure where to put them all. Water new transplants thoroughly and keep them well watered until they have re-established themselves.

The larger the sections of the division, the more quickly the plant will re-establish itself and grow to blooming size again. For example, a perennial divided into four sections will bloom sooner than the same one divided into ten sections. Very small divisions may benefit from being planted in pots until they are bigger and better able to fend for themselves in the border.

Newly planted divisions will need extra care and attention when they are first planted. They will need regular watering and, for the first few

are several signs that a perennial should be divided:
• the center of the plant has died out
• the plant is no longer flowering as profusely as it did in previous years
• the plant is encroaching on the growing space of other plants sharing the bed

It is relatively easy to divide perennials. Begin by digging up the entire clump and knocking any large clods of soil away from the root ball. The clump can then be split into several pieces. A small plant with fibrous roots can be torn into sections by hand. A large plant can be

days, shade from direct sunlight. A light covering of burlap or damp newspaper should be sufficient to shelter them for this short period. Divisions that have been planted in pots should be moved to a shaded location.

There is some debate about the best time to divide perennials. Some gardeners prefer to divide perennials while they are dormant, whereas others feel perennials establish themselves more quickly if divided when they are growing vigorously. You may wish to experiment with dividing at different times of the year to see what works best for you. There are some general guidelines for dividing perennials based on when they bloom. Divide spring-blooming perennials in very early spring, after they bloom, or in early fall. Divide fall-blooming perennials in spring. Divide summer-blooming perennials in early spring or after they bloom. Of course, there are always exceptions, but that is where the experimentation come in.

There are some perennials that should not be divided, either because they have a single crown, or because the roots do not like being disturbed. Others are content to remain in one spot for a long time.

Perennials that Should not be Divided

Baptisia
Columbine
Euphorbia (some species)
Gas plant
Monkshood
Peony

Peony (above), False indigo (below)

Problems and Pests

Perennial gardens are both a liability and an asset when it comes to pests and diseases. Perennial beds often contain a mixture of different plant species, and many insects and diseases attack only one species of plant. Mixed beds make it difficult for pests and diseases to find their preferred hosts and establish a population. At the same time, because the plants are in the same spot for many years, the problems can become permanent. The advantage is that the beneficial insects, birds and other pest-devouring organisms can also develop permanent populations.

For many years pest control meant spraying or dusting. The goal was to try to eliminate every insect—the good and the bad—in the garden. A more moderate approach is advocated today. The goal is now to maintain problems at levels at which only negligible damage is done. Chemicals can endanger the gardener and his or her family, and they kill the good organisms as well as the bad ones, leaving the garden open to even worse attacks.

There are four steps in managing pests organically. The cultural controls are the most important. The physical controls come next, followed by the biological controls. The chemical controls should only be used when the first three possibilities have been exhausted.

Cultural controls are the regular gardening techniques you use in the day-to-day care of your garden. Growing perennials in the conditions they prefer and keeping your soil healthy with plenty of organic matter are just two of the cultural controls you can use to keep pests manageable in your garden. Choose resistant varieties of perennials that are not prone to problems. Space perennials so that they have good air circulation around them and are not stressed from competing for light, nutrients and room. Remove plants from the garden if they are constantly decimated by the same pests every year. Remove and destroy diseased foliage and prevent the spread of disease by keeping your gardening tools clean and by tidying up fallen leaves and dead plant matter at the end of the growing season.

Physically removing weeds and pests is an effective control strategy.

Physical controls are generally used to combat insect problems. These include things like picking the insects off the perennials by hand, which is not as daunting a solution as it seems if you catch the problem when it is just beginning. Other physical controls are barriers that stop the insects from getting to the plant or traps that either catch or confuse the insects. The physical control of diseases can generally only be accomplished by removing of the infected perennial parts to prevent the spread of the problem.

Biological controls make use of the populations of natural predators. Animals including birds, snakes, frogs, spiders, lady beetles and certain bacteria can play a role in keeping pest populations at a manageable level. Encourage these creatures to take up permanent residence in your garden. A birdbath and birdfeeder will encourage birds to enjoy your yard

and feed on a wide variety of insect pests. Many beneficial insects are probably already living in your garden and they can be encouraged to stay with alternate food sources. Beneficial insects also eat the nectar from flowers. The flowers of nectar plants like yarrow are popular with many predatory insects.

In a perennial garden you should rarely have to resort to chemicals, but if it does become necessary there are some organic options available. Organic sprays are no less dangerous than synthetic chemical ones, but they will break down into harmless

Frogs eat many insect pests.

compounds because they come from natural sources. The main drawback to using any chemicals is that they may also kill the beneficial insects you have been trying to attract to your garden. Organic pest controls should be available at local garden centers and should be applied at the rates and for the pests recommended on the packages. Proper and early identification of problems is vital for finding a quick solution.

Whereas cultural, physical, biological and chemical controls are all possible defenses against insects, disease must be controlled culturally. Prevention is often the only hope: once a plant has been infected, it should probably be destroyed, so as to prevent the spread of the disease.

Glossary of Pests & Diseases

Aphids

Adult ladybird beetle

Aphids

Tiny, pear-shaped insects, winged or wingless; green, black, brown, red or gray. Cluster along stems, on buds and on leaves. Suck sap from plants; cause distorted or stunted growth. Sticky honeydew forms on surfaces and encourages sooty mold growth. Woolly adelgids are a type of aphid.

What to Do: Squish small colonies by hand; dislodge with brisk water spray; encourage predatory insects and birds that feed on aphids; spray serious infestations with insecticidal soap or neem oil according to package directions.

Aster Yellows

see Viruses

Beetles

Many types and sizes; usually round with hard, shell-like outer wings covering membranous inner wings. Some are beneficial, e.g., ladybird beetles ("ladybugs"); others, e.g., Japanese beetles, leaf skeletonizers and weevils, eat plants. Larvae: see Borers, Grubs. Leave wide range of chewing damage: make small or large holes in or around margins of leaves; consume entire leaves or areas between leaf veins ('skeletonize'); may also chew holes in flowers. Some bark beetle species carry deadly plant diseases.

What to Do: Pick beetles off at night and drop them into an old coffee can half filled with soapy water (soap prevents them from floating and climbing out).

Blight

Fungal diseases, many types; e.g., leaf blight, needle blight, snow blight. Leaves, stems and flowers blacken, rot and die.

What to Do: Thin stems to improve air circulation; keep mulch away from base of plants; remove debris from garden at end of growing season. Remove and destroy infected plant parts.

Bugs (True Bugs)

Small insects, up to $1/2$" long; green, brown, black or brightly colored and patterned. Many beneficial; a few pests, such as lace bugs, pierce plants to suck out sap. Toxins may be injected that deform plants; sunken areas left where pierced; leaves rip as they grow; leaves, buds and new growth may be dwarfed and deformed.

What to Do: Remove debris and weeds from around plants in fall to destroy overwintering sites. Spray plants with insecticidal soap or neem oil according to package directions.

Canada Geese

If you have a pond on your property, you may find flocks of geese nesting and making themselves at home in your garden. An adult goose drops 2 lbs of fecal matter daily, which can present health problems, contaminate your pond water. The geese can ravage your plants and, specifically, grasses.

What to Do: There are several humane ways to discourage the geese: using repellants, windmills, kites, sonar devices, and others recommended by the Humane Society of the U.S. These have been compiled in a guide they will mail to you at your request. Don't feed the geese. Plant tall fescue instead of tender Kentucky bluegrass, and use unpalatable groundcovers such as pachysandra. Place large stones, tall grass or a short fence around the pond's edge to keep the geese from stepping up. Plant hedges to reduce the goose's ability to detect predators. Discourage geese from

Lygus bug (above), Canada geese (below)

walking between ponds and feeding areas; a low barrier formed by string tied at a height of about 4" above the ground between several stakes will stop them from moving from the water to areas around the pond.

Case Bearers

see Caterpillars

Caterpillars

Larvae of butterflies, moths, sawflies. Include bagworms, budworms, case bearers, cutworms, leaf rollers, leaf tiers, loopers. Chew foliage and buds; can completely defoliate a plant if infestation severe.

What to Do: Removal from plant is best control. Use high-pressure water and soap or pick caterpillars off small plants by hand. Control biologically using the naturally occurring soil bacterium *Bacillus thuringiensis* var. *kurstaki* or *B.t.* (commercially available), which breaks down gut lining of caterpillars.

Deer

An estimated 1.6 to 2 million white-tailed deer roam loose in Pennsylvania and present a major problem for gardeners, not to mention drivers. Each adult deer eats 5–7 lbs of food daily and needs 10–12 acres on which to browse. Beautiful as they are, they are capable of decimating crops, woodlands and pretty much everything they can get their lips around, including the plants that deer are not supposed to like. When they're hungry, they're not picky eaters. They also like to rub their antlers on sapling trees and can easily kill them by girdling the bark and snapping the poor little trees in two. Deer also host ticks that carry Lyme disease and Rocky Mountain spotted fever.

What to Do: There are many deterrents that work for a while: encircle immature shrubs with tall upright sticks; spray any number of deer repellents on plants; strew fabric strips scented with creosote, gasoline or turpentine over shrubbery; place rotted meat, dangling soap bars, mothballs or blood meal around the garden; sprinkle all manner of wild animal excretions around as repellants; squirt pepper sprays on plants to make them distasteful; use noisemaking devices or water spritzers to startle deer; mount flashy aluminum or moving devices here and there to scare them away. But, in the end, the only surefire prevention is to enclose your property with a very tall and sturdy deer fence, which is sometimes impractical and may make you feel as though you have committed a crime and are being incarcerated. Some problems we end up having to learn to live with as best we can.

Galls

Unusual swellings of plant tissues that may be caused by insects or diseases. Can affect leaves, buds, stems, flowers, fruit. Often a specific gall affects a single genus or species.

What to Do: Cut galls out of plant and destroy them. Galls caused by insects usually contain the insect's eggs and juvenile stages. Prevent such galls by controlling insect before it lays eggs; otherwise try to

Caterpillar eating flowers

remove and destroy infected tissue before young insects emerge. Generally insect galls are more unsightly than damaging to plant. Galls caused by diseases often require destruction of plant. Avoid placing other plants susceptible to same disease in that location.

Grubs

Larvae of different beetles, commonly found below soil level; usually curled in a C-shape. Body white or gray; head may be white, gray, brown or reddish. Problematic in lawns; may feed on roots of perennials. Plant wilts despite regular watering; may pull easily out of ground in severe cases.

What to Do: Toss any grubs found while digging onto a stone path, driveway, road or patio for birds to devour; apply parasitic nematodes or milky spore to infested soil (ask at your local garden center).

Leafhoppers & Treehoppers

Small, wedge-shaped insects; can be green, brown, gray or multicolored. Jump around frantically when disturbed. Suck juice from plant leaves, cause distorted growth, carry diseases such as aster yellows.

What to Do: Encourage predators by planting nectar-producing species like yarrow. Wash insects off with strong spray of water; spray with insecticidal soap or neem oil according to package directions.

Leaf Miners

Tiny, stubby larvae of some butterflies and moths; may be yellow or green. Tunnel within leaves leaving winding trails; tunneled areas lighter in color than rest of leaf. Unsightly rather than major health risk to plant.

What to Do: Remove debris from area in fall to destroy overwintering sites; attract parasitic wasps with nectar plants such as yarrow. Remove and destroy infected foliage; can sometimes squish by hand within leaf.

Leaf miner galleries in leaf

Leaf Rollers
see Caterpillars

Leaf Skeletonizers
see Beetles

Leaf Spot

Two common types: one caused by bacteria and the other by fungi. *Bacterial:* small brown or purple spots grow to encompass entire leaves; leaves may drop. *Fungal:* black, brown or yellow spots; leaves wither; e.g., scab, tar spot.

What to Do: Bacterial infection more severe; must remove entire plant. For fungal infection, remove and destroy infected plant parts. Sterilize removal tools; avoid wetting foliage or touching wet foliage; remove and destroy debris at end of growing season.

Mealybugs

Tiny crawling insects related to aphids; appear to be covered with white fuzz or flour. Sucking damage stunts and stresses plant. Excrete honeydew that promotes growth of sooty mold.

What to Do: Remove by hand from smaller plants; wash plant off with soap and water; wipe off with alcohol-soaked swabs; remove heavily infested leaves; encourage or introduce natural predators such as mealybug destroyer beetle and parasitic wasps; spray with insecticidal soap. Keep in mind larvae of mealybug destroyer beetles look like very large mealybugs.

Mice

Mice love to burrow under mulch in the winter and enjoy chewing the roots of your plants, any tree bark they can reach, tulip bulbs and many other underground goodies. Even plants or roots stored in cool porches, garages or sheds can be killed or damaged.

What to Do: Fine wire mesh can prevent mice from getting at your plants in winter, though they are quite ingenious and may find their way through or around any barrier you erect. Bulbs and lifted roots can be rolled in talcum powder, garlic powder or bulb protectant spray before storing or planting. One of the best defenses against destructive mice is having a cat. (Borrow or encourage your neighbor's, if you must.)

Mildew

Two types, both caused by fungus, but with slightly different symptoms. *Downy mildew:* yellow spots on upper sides of leaves and downy fuzz on undersides; fuzz may be yellow, white or gray. *Powdery mildew:* white or gray powdery coating on leaf surfaces, doesn't brush off.

What to Do: Choose resistant cultivars; space plants well; thin stems to encourage air circulation; tidy any debris in fall. Remove and destroy infected leaves or other parts.

Powdery mildew

Mites

Tiny, eight-legged relatives of spiders; do not eat insects, but may spin webs. Almost invisible to naked eye; red, yellow or green; usually found on undersides of plant leaves. Examples: bud mites, spider mites, spruce mites. Suck juice out of leaves. May see fine webbing on leaves and stems; may see mites moving on leaf undersides; leaves become discolored and speckled, then turn brown and shrivel up.

What to Do: Wash off with strong spray of water daily until all signs of infestation are gone; predatory mites available through garden centers; spray plants with insecticidal soap.

Moles and Gophers

These critters burrow under the soil making tunnels throughout your property in search of insects, grubs and earthworms. Unfortunately, the tunnels can create runways for voles, who will eat your prize plants from below ground.

What to Do: Castor oil spilled down the mole's runway is effective. This is the primary ingredient in most repellants made to thwart moles and gophers. You can buy it in granulated pellet form, too. Noisemakers and predator urine are also used. Humane trapping is also good. Having a cat or dog to guard the property is effective, too.

Mosaic

see Viruses

Nematodes

Tiny worms that give plants disease symptoms. One type infects foliage and stems; the other infects roots. *Foliar:* yellow spots that turn brown on leaves; leaves shrivel and wither; problem starts on lowest leaves and works up plant.

Root-knot: plant is stunted, may wilt; yellow spots on leaves; roots have tiny bumps or knots.

What to Do: Mulch soil, add organic matter, clean up debris in fall, don't touch wet foliage of infected plants. Can add parasitic nematodes to soil. Remove infected plants in extreme cases.

Rabbits

Peter Rabbit doesn't only hang around Mr. McGregor's garden. When you find the heads missing from your tulips, chances are bunnies have been around. They will eat as much of your garden as deer, and munch on the bark of your trees and shrubs, too.

What to Do: The same deterrents that work for deer will usually keep rabbits away, as will humane trapping. Having a cat or dog patrol your garden may also be effective.

Rot

Several different fungi that affect different parts of plant and can kill plant. *Crown rot:* affects base of plant, causing stems to

blacken and fall over and leaves to yellow and wilt. *Root rot:* leaves yellow and plant wilts; digging up plant shows roots rotted away. White rot: a "watery decay fungus" that affects any part of plant; cell walls appear to break down, releasing fluids.

What to Do: Keep soil well drained; don't damage plant if you are digging around it; keep mulches away from plant base. Destroy infected plant if whole plant affected.

Raccoons

Raccoons are especially fond of devouring any fruit and some vegetables you are cultivating. They can carry rabies and canine distemper. They also eat grubs, insects and mice, so they're sometimes helpful to gardeners.

What to Do: Don't allow access to trashcans or pet food; humane traps and relocation are the best solutions. Call your local SPCA or Humane Society to relocate them.

Rust

Fungi. Pale spots on upper leaf surfaces; orange, fuzzy or dusty spots on leaf undersides. Examples: blister rust, hollyhock rust.

What to Do: Choose rust-resistant varieties and cultivars; avoid handling wet leaves; provide plant with good air circulation; clear up garden debris at end of season. Remove and destroy infected plant parts.

Scab

see Leaf Spot

Scale Insects

Tiny, shelled insects that suck sap, weakening and possibly killing plant or making it vulnerable to other problems. Once female scale insect has pierced plant with mouthpart, it is there for life. Juvenile scale insects called crawlers.

What to Do: Wipe off with alcohol-soaked swabs; spray with water to dislodge crawlers; prune out heavily infested branches; encourage natural predators and parasites; spray dormant oil in spring before bud break.

Slugs

Mollusks up to 8" long, many smaller. Slimy, smooth skin; gray, green, black, beige, yellow or spotted. Leave large, ragged holes in leaves and silvery slime trails on and around plants.

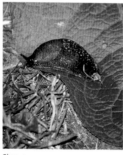
Slug

What to Do: Attach strips of copper to wood around raised beds or to smaller boards inserted around susceptible groups of plants; slugs get shocked if they try to cross copper surfaces. Pick off by hand in the evening and squish with boot or drop in can of soapy water. Spread wood ash or diatomaceous earth (available in garden centers) on ground around plants; it will pierce their soft bodies and dehydrate them. Slug baits containing iron phosphate (also available in garden centers) are not harmful to humans or animals and control slugs very well when used according to package directions. If slugs damaged garden last season, begin controls as soon as new green shoots appear in spring.

Squirrels

These entertaining critters unearth and eat bulbs and corms, as well as flowers, fruits and vegetables. They chew on sugar maples to satisfy their sweet tooth and hone their other teeth on almost everything else. They raid the birdfeeder and often eat the feeder itself. Often, they bury their food for later consumption, each meal stored in a separate larder, which can cover a fair amount of your property and result in seeds germinating and plants springing up where you never wanted them. Squirrels are sometimes harder to outwit than deer. After all, deer can't climb your downspouts, chew their way into your attic and eat your electrical wiring.

What to Do: To keep squirrels from digging around your garden and in flower pots, you can cut heavy metal screening with a ½" square grid (hardware cloth) to fit around the stem of a plant. Caging entire plants is effective, if you don't mind your garden looking like a zoo. Removing enticing food supplies is effective, but often impractical. Trapping is the only way to remove squirrels. After humanely snaring in a Havahart trap (havahart.com), you can drive them over the border to some state you dislike, but usually other squirrels soon move in.

Sooty Mold

Fungus. Thin black film forms on leaf surfaces and reduces amount of light getting to leaf surfaces.

What to Do: Wipe mold off leaf surfaces; control insects like aphids, mealybugs, whiteflies (honeydew left on leaves encourages mold).

Tar Spot

see Leaf Spot

Thrips

Tiny, slender insects, difficult to see; may be visible if you disturb them by blowing gently on an infested flower. Yellow, black or brown; narrow, fringed wings. Suck juice out of plant cells, particularly in flowers and buds, causing mottled petals and leaves, dying buds, distorted and stunted growth.

What to Do: Remove and destroy infected plant parts; encourage native predatory insects with nectar plants like yarrow; spray severe infestations with insecticidal soap or neem oil according to package directions.

Viruses

Plant may be stunted and leaves and flowers distorted, streaked or discolored. Examples: aster yellows, mosaic virus, ringspot virus.

What to Do: Viral diseases in plants cannot be treated. Control insects, such as aphids, leafhoppers and whiteflies, that spread disease. Destroy infected plants.

Mosiac virus

Voles

Mouse-like creatures, voles usually damage plants at or just beneath the soil surface. Voles are mostly herbivorous, feeding on a variety of grasses, vegetables, herbaceous plants, bulbs (lilies are a favorite) and tubers. They eat bark and roots of trees, usually in fall or winter. They store seeds and other plant matter in underground chambers.

What to Do: Wire fences at least 12" above the ground with a mesh size of ¹/₄" or less can help exclude voles from entire gardens. These fences can either stand alone or be attached to the bottom of an existing fence. Bury the bottom edge of the fence 6–10" deep to prevent voles from tunneling beneath it. A weed-free barrier on the outside of the fence will increase its effectiveness. Burrow fumigants are not effective because the vole's burrow system is shallow and has many open holes. Electromagnetic or ultrasonic devices and flooding are also ineffective against voles. When voles are not numerous or when the population is concentrated in a small area, trapping may be effective. Use a sufficient number of traps to control the population: for a small garden a dozen traps is probably the minimum number required, and for larger areas at least 50 or more may be needed. Again, having a dog or cat is a deterrent. Be certain not to use poisonous repellents or baits if you have pets or little children romping around the garden.

Weevils
see Beetles

Whiteflies
Flying insects that flutter up into the air when plant is disturbed. Tiny, moth-like, white; live on undersides of plant leaves. Suck juice out of leaves, causing yellowed leaves and weakened plants; leave behind sticky honeydew on leaves, encouraging sooty mold growth.

What to Do: Destroy weeds where insects may live. Attract native predatory beetles and parasitic wasps with nectar plants like yarrow; spray severe cases with insecticidal soap. Can make a sticky flypaper-like trap by mounting tin can on stake; wrap can with yellow paper and cover with clear plastic bag smeared with petroleum

jelly; replace bag when covered in flies.

Wilt
If watering doesn't help wilted plant, one of two wilt fungi may be at fault. *Fusarium* wilt: plant wilts, leaves turn yellow then die; symptoms generally appear first on one part of plant before spreading to other parts. *Verticillium* wilt: plant wilts; leaves curl up at edges; leaves turn yellow then drop off; plant may die.

What to Do: Both wilts difficult to control. Choose resistant plant varieties and cultivars; clean up debris at end of growing season. Destroy infected plants; solarize (sterilize) soil before replanting (may help if entire bed of plants lost to these fungi)—contact local garden center for assistance.

Woolly Adelgids
see Aphids

Worms
see Caterpillars, Nematodes

You can make your own insecticidal soap at home.

Mix 1 tsp. of mild dish detergent or pure soap (biodegradable options are available) with 1 qt. of water in a clean spray bottle.

Spray the surfaces of your plants and rinse well within an hour of spraying to avoid foliage discoloration.

Gardens to Visit in Pennsylvania

AMBLER
Temple University at
Ambler—Landscape
Arboretum
(215) 283-1292
www.temple.edu/ambler/la-
hort/aboutarboretum.htm

CENTRE HALL
Rhoneymeade Arboretum
and Sculpture Garden
(814) 364-1527
rhoneymeade-usa.org

CHADDS FORD
Brandywine Conservancy/
Brandywine River
Museum—Gardens
(610) 388-8327
brandywinemuseum.org/
gardens.html

ERIE
Erie Zoological Park and
Botanical Gardens of
Northwestern Pennsylvania
(814) 864-4091
eriezoo.org/discover_gar-
dens.htm

HERSHEY
Hershey Gardens
(717) 534-3492
hersheygardens.org

KENNETT SQUARE
Longwood Gardens
(610) 388-1000
longwoodgardens.org

MEADOWBROOK
Meadowbrook Farm
(215) 887-5900
meadowbrook-farm.com

PHILADELPHIA
Bartram's Garden
(215) 729-5281
bartramsgarden.org

Morris Arboretum of the
University of Pennsylvania
(215) 247-5777
www.upenn.edu/arboretum

PITTSBURGH
Phipps Conservatory and
Botanical Gardens
(412) 622-6914
phipps.conservatory.org

SWARTHMORE
Scott Arboretum of Swarthmore College
(610) 328-8025
scottarboretum.org

UNIVERSITY PARK
The Pennsylvania State University—The Arboretum at
Penn State
(814) 865-9118
arboretum.psu.edu

The Pennsylvania State
University—Horticultural
Trial Gardens
(814) 863-7725
hortweb.cas.psu.edu/
research/trial.html

WAYNE
Chanticleer
(610) 687-4163
www.chanticleergarden.org

Garden Shows in Pennsylvania

**Spring and Fall Home
and Garden Show**, in
Wlikes-Barre, Philadelphia,
Fort Washington, Reading,
Pittsburgh, Allentown and
other parts of the state

**Philadelphia Flower
Show**, annual, early March

**Lehigh Valley Flower
Show**, Allentown

**Mid-Atlantic Flower
Show**, annual, spring,
Susquehanna Valley in York

There are many smaller flower
shows to visit. Most of the
arboreta and public gardens
throughout the state have
exemplary perennial displays
for inspiration.

About This Guide

The perennials in this book are organized alphabetically by their most familiar common names, which in some cases is also the proper botanical name. If you are familiar only with the common name for a plant, you should be able to find the plant easily in the book. The botanical name is always listed in italics below the common name. We strongly encourage you to learn these botanical names. Only the botanical name for a plant defines exactly what plant it is, everywhere on the planet and for any language. Learning and using the botanical names for your plants allows you to discuss, research and purchase plants with confidence and accuracy.

At the beginning of each entry are height and spread ranges, which include the measurements for all recommended species and varieties, along with the full range of flower colors for those plants that produce flowers. Also included at the top of each entry is zone information, which indicates the overall hardiness of a particular genus. Specific species with different hardiness zone ranges are identified in the Recommended section of each entry. At the back of the book, the **Quick Reference Chart** (p. 340) summarizes different features and requirements of the perennials; you will find this chart handy when planning for diversity in your garden.

Each entry gives clear instructions and tips for seeding, planting and growing the perennial, and it recommends some of our favorite species and varieties. Keep in mind that many more hybrids, cultivars and varieties are often available than we have space to mention. Check with local greenhouses or garden centers when making your selection. Many perennials are available that we did not include in this book, and we encourage you to explore. That said, we present plenty of wonderful perennials in this book to provide you with many seasons of gardening pleasure.

Pests or diseases that commonly affect perennials are also listed for each entry. Consult the **Problems & Pests** (p. 56) section of the introduction for information on how to solve these problems.

Finally, we have kept jargon to a minimum, but check the **Glossary** on p. 346 for any unfamiliar terms.

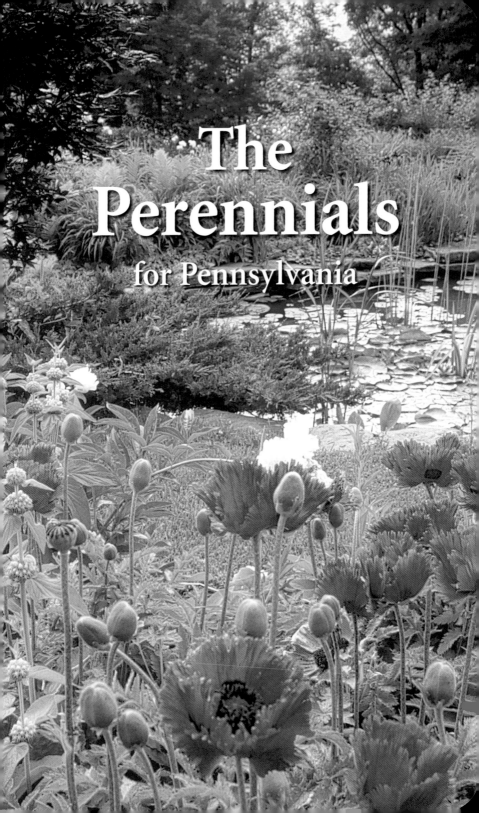

The
Perennials
for Pennsylvania

Ajuga
Ajuga

Also Called: Bugleweed

Height: 3–12" **Spread:** 6–36" **Flower color:** purple, blue, pink, white; plant grown for foliage **Blooms:** late spring to early summer **Zones:** 3–8

ONCE A STAPLE OF MEDICINAL HERB GARDENS, THIS MEMBER OF the boisterous mint family is now more commonly relegated to the lowly status of ground creeper. But don't let that dissuade you from trying one of the stunning few out of the 40 or 50 species available to us. *A.reptans* can be invasive, but few people waste time with that ho-hum species when they can grow some of the colorful cultivars and other species. *A. pyramidalis* 'Metallica Crispa,' for instance, with almost-black, crinkled, lustrous foliage and tiny, deep blue flowers, like most of its new cousins, is quite tame, forming well-mannered mats to set off other sun or shade-loving plants.

According to European folk myths, ajugas could cause fires if brought into the house.

Planting

Seeding: Not recommended; foliage often reverts to green in seedlings

Planting out: Spring, summer or fall

Spacing: 12–18"

Growing

Ajugas develop the best leaf color in **partial** or **light shade** but tolerate full shade. The leaves may become scorched when exposed to too much sun. Any **well-drained soil** is suitable. Winter protection, such as evergreen branches laid across the plants, is recommended if snow cover isn't dependable in your garden. Divide these vigorous plants any time during the growing season.

Remove any new growth or seedlings that don't show the hybrid leaf coloring.

Tips

Ajugas make excellent groundcovers for difficult sites, such as exposed slopes and dense shade. They are

A. tenorii 'Chocolate Chip' (above), *A. reptans* 'Caitlin's Giant' (below)

A. reptans cultivar (above), *A. reptans* (below)

also attractive groundcovers in shrub borders, where their dense growth will prevent the spread of all but the most tenacious weeds.

If you plant ajugas next to a lawn, you may soon be calling them weeds. Because they spread readily by stolons (creeping, above-ground shoots), these plants can easily invade a lawn, and their low growth escapes mower blades. The spread of ajugas may be somewhat controlled by the use of bed-edging materials. If an ajuga does start to take over, however, it is easy to rip out, and the soil it leaves behind will be soft and loose from the penetrating roots. Use an ajuga as a scout plant to send ahead and prepare the soil before you plant anything fussier in a shaded or woodland garden.

If you're not obsessive about edging beds, plant these aggressive growers in areas bordered by brick or cement. Close spacing and regular watering helps these plants spread quickly and fill in, preventing weeds from springing up among the groundcover.

Recommended

A. pyramidalis (*A. metallica*; pyramid bugleweed) is a low-growing, clump forming plant. New clumps form along the rhizomes. Plants grow 3–5" tall and spread 6–10" in three years. These plants prefer a shady, moist spot in the garden. **'Leprechaun'** is a small, 3–4" plant with very crinkly, dark green leaves. **'Metallica Crispa'** is a very slow-growing plant with bronzy, crinkly foliage. The violet blue flowers contrast beautifully with the foliage.

A. reptans is the best-known species. It is a low, quick-spreading ground-cover that grows about 6" tall and spreads 18–24". It can be very invasive and is rarely grown in favor of the more civilized cultivars. **'Burgundy Glow'** has variegated foliage in shades of bronze, green, white and pink. The habit is dense and compact. **'Caitlin's Giant'** has large, bronze leaves. It bears short spikes of bright blue flowers in spring. **'Purple Brocade'** has dark purple leaves that turn nearly black in winter. Spikes of dark blue flowers are produced in spring.

A. tenorii **'Chocolate Chip'** is a low, creeping plant 4–6" tall and 12–36" in spread, with chocolaty bronze, teardrop-shaped leaves. It bears spikes of blue flowers in early summer.

Problems & Pests

Occasional problems with crown rot, leaf spot and root rot can be

A. reptans cultivar (above), *A. reptans* (below)

avoided by providing good air circulation and by ensuring the plant is not standing in water for extended periods.

Ajuga reptans *is widely used in homeopathic remedies for throat and mouth irritations.*

Allium

Allium

Also Called: Ornamental Onion, Flowering Onion
Height: 1–6' **Spread:** 4–24" **Flower color:** green, pink, purple, red, blue, white **Blooms:** summer **Zones:** 3–9

ALTHOUGH CULINARY ALLIUMS (ONIONS, SHALLOTS, GARLIC, LEEKS and chives) have been used as antiseptics, aphrodisiacs, standard vampire repellants and treatments for everything from freckles to smallpox since 3500 BC, not until the mid-19th century were any used ornamentally. If *Allium vineale* and *A. canadense*—the wild garlic and onions that plague our lawns—alarm you, avoid the alluring, self-seeding garden gangster *A. tuberosum* (garlic chives). Another caveat: some, like beautiful, wine-colored drumstick alliums (*A. sphaerocephalon*) unfortunately emerge so closely resembling those wild garlic and onion lawn brats that every fall I plant some only to inadvertently "weed" them out before they ever bloom.

Alliums go well with old roses, and advocates of companion planting suggest garlic grown among roses protects against blackspot and powdery mildew. It's also worthwhile to note that trouble-making deer, squirrels and raccoons are often repulsed by them, possibly because animal etiquette dictates that approaching a potential mate with allium breath invites certain rejection.

Planting

Seeding: Spring

Planting out: Fall or spring

Spacing: 6–18"

Growing

Alliums grow best in **full sun**. The soil should be **average to fertile, moist and well drained.** Bulbs are best planted in fall, but if available, plants can be planted out in spring. Plant bulbs 2–4" deep, depending on the size of the bulb; larger bulbs should be planted more deeply than smaller ones.

Tips

Alliums are best planted in groups in beds or borders where they can be left to naturalize. Most will self-seed freely when left to their own devices. The foliage tends to die back as the plants come into bloom, so low-growing companion plants are a good idea to give the planting a fuller, more attractive appearance.

Recommended

A. caeruleum (blue globe onion) has fairly short, narrow leaves. Plants grow about 24" tall and spread 4–6". Dense clusters of bright blue flowers are produced on stiff stems in early summer.

A. sphaerocephalon (opposite), *A. giganteum* (above)

A. giganteum (giant onion) has large, strap-like leaves. Plants grow up to 6' tall and spread 12–24". Dense, globe-shaped clusters of purpley pink flowers are borne in summer. 'Globemaster' bears rounded clusters of purple flowers. It grows about 32" tall.

A. schubertii has bright green, strap-like leaves. Plants grow 12–24" tall and spread 8–12". Bears light purple flowers that resemble exploding fireworks or 4th of July sparklers. The dried seedheads are wonderful for arrangements if you can catch them before they roll away like tumbleweeds.

A. sphaerocephalon (drumstick allium, round-headed garlic) has long, narrow leaves. Plants grow up to 36" tall and spread 12–18". Dense clusters of green, pink or dark red flowers are borne in summer. Small bulbs, or bulbils, may form on the flower heads.

Problems & Pests

Bulbs may rot in soil that is wet or poorly drained. Problems with fungal leaf spots, rust, mildew, white rot, smut, onion flies and thrips can also occur.

Anemone

Anemone

Also Called: Windflower

Height: 2"–5' **Spread:** 6–24" **Flower color:** white, pink, blue, lilac **Blooms:** spring, summer, early fall **Zones:** 5–8

A DIVERSE GENUS OF SOME 100 SPECIES, ANEMONE PROVIDES enough flowers to blow about in the garden from early spring through late fall. Grecian windflowers brightly carpet the early spring garden floor with vivid blue, lavender, pink and/or white, daisy-like flowers that multiply year after year. The shade-loving wood anemones are early, too. They are a few inches taller, and offer softer blue, pink, white, mauve and double blossoms. Tallest of all are the fall-blooming Japanese hybrids with single, double and semi-double forms in rose, pink, white and lilac.

Planting

Seeding: Not recommended

Planting out: Spring

Spacing: 4–18"

Growing

Anemones prefer **partial** or **light shade** but tolerate full sun with adequate moisture. The soil should be of **average to high fertility, humus rich and moist.** Grecian windflower prefers a light, sandy soil. Meadow anemone needs regular watering when first planted in order to become established. While dormant, anemones should have dry soil. Mulch Japanese anemone the first winter to help it establish.

Divide Grecian windflower in summer and other anemones in spring or fall.

Deadheading will only keep plants tidy because removing spent flowers does not extend the bloom. A few anemones produce small, fluffball-like seedheads, which add another point of interest to the garden.

Tips

Anemones make beautiful additions to lightly shaded borders, woodland gardens and rock gardens. Species that go dormant in mid-summer make good companions for plants that tend to sprout late in spring and fill in.

Recommended

A. *blanda* (Grecian windflower) is a low, spreading, tuberous species that bears blue flowers in spring. It grows 6–8" tall, with an equal

A. blanda 'White Splendor' (above) *A. nemerosa* (below)

A. x *hybrida* 'Max Vogel'

A. nemorosa (center & below)

spread. **'Blue Star'** has pale blue flowers. **'Pink Star'** has pink flowers with yellow centers. **'White Splendor'** is a vigorous plant with white flowers.

A. **x** *hybrida* (Japanese anemone, hybrid anemone) is an upright, suckering hybrid. It grows 2–5' tall, spreads about 24" and bears pink or white flowers from late summer to early fall. Many cultivars are available. **'Honorine Jobert,'** one of the oldest cultivars, has plentiful white flowers. **'Max Vogel'** has large, pink flowers. **'Pamina'** has pinkish red, double flowers. **'Whirlwind'** has white, semi-double flowers.

A. nemorosa (wood anemone) is a low, creeping perennial that grows 3–10"

tall and spreads 12" or more. The spring flowers are white, often flushed with pink. **'Flore Pleno'** has double, white flowers. **'Rosea'** has red-purple flowers.

A. quinquefolia (wood anemone) is a low-growing, spreading plant that grows 2–8" tall and spreads 12–24". Pink flushed white flowers are borne in mid- to late spring. Plants often die back, going dormant in mid-summer.

Problems & Pests

Rare but possible problems include leaf gall, downy mildew, smut, fungal leaf spot, powdery mildew, rust, nematodes, caterpillars, slugs and flea beetles.

A. x *hybrida* 'Whirlwind' (above)
A. x *hybrida* (below)

Artemisia
Artemisia

Also Called: Wormwood, Sage, Dusty Miller
Height: 6"–6' **Spread:** 12–36" **Flower color:** white or yellow, generally inconspicuous; plant grown for foliage **Blooms:** late summer, mid-fall
Zones: 3–8

A VIGOROUS, DEER-PROOF GROWER IDEALLY SUITED TO HOT, DRY, sunny exposures and poor, rocky soils, artemisia has a toughness that is at odds with its lacy, delicate appearance. Its silver-gray or blue-green foliage makes an indispensable accompaniment for plants with dark leaves and flowers, which tend to recede in a border, and foils jarring color combinations in the garden. Artemesia is also beautiful in its own right, and a mainstay for gardens with a white theme or those to be enjoyed at night. Use this plant around a pool to help illuminate evening swim parties. The leaves are also pleasantly aromatic, especially when crushed.

There are almost 300 species of Artemisia distributed around the world.

Planting

Seeding: Not recommended

Planting out: Spring, summer or fall

Spacing: 10–36"

Growing

Artemisias grow best in **full sun**. The soil should be of **low to average fertility and well drained**. These plants dislike wet, humid conditions.

Artemisias respond well to pruning in late spring. Whenever they begin to look straggly, they can be cut back hard to encourage new growth and to maintain a neater form. Divide every one or two years when plants appear to be thinning in the centers.

Tips

Use artemisias in rock gardens and borders. Their silvery gray foliage makes them good backdrop plants to use behind brightly colored flowers. They are also useful for filling in spaces between other plants. Smaller forms may be used to create knot gardens.

Some species can become invasive. If you want to control horizontal spreading of a rhizomatous artemisia, plant it in a bottomless container. Sunk into the ground, the hidden container prevents the plant from spreading beyond the container's edges. You can maintain good drainage by removing the bottom of the container.

A. ludoviciana cultivar (above), *A.* 'Powis Castle' (below)

A. lactiflora (above), *A. ludoviciana* cultivar (below)

Recommended

*A. **absinthium*** (common wormwood) is a clump-forming, woody-based perennial 24–36" tall and about 24" in spread. It has aromatic, hairy gray foliage and bears inconspicuous yellow flowers in late summer. **'Lambrook Silver'** has attractive, silver-gray foliage. (Zones 4–8)

*A. **lactiflora*** (white mugwort) is an upright, clump-forming species 4–6' tall and 24–36" in spread. It is one of the few artemisias to bear showy flowers; its attractive, creamy white blooms appear from late summer to mid-fall. The foliage of this hardy species is dark green or gray-green. **'Guizho'** is an outstanding cultivar with dark green, almost black, foliage and reddish stems that contrast well with its creamy flowers. (Zones 3–8)

*A. **ludoviciana*** (white sage, silver sage) is an upright, clump-forming species. It grows 2–4' tall and spreads 24". The foliage is silvery white and the flowers are inconspicuous. The species is not grown as often as the cultivars. **'Valerie Finnis'** is a good choice for hot, dry areas. It has very wide, silvery leaves, is less invasive than the species and combines beautifully with many other perennials. (Zones 4–8)

A. **'Powis Castle'** is compact, mounding and shrubby, reaching 24–

A. absinthium *lent its name and its flavor to the liqueur absinthe, which was once popular in France. Absinthe containing the wormwood toxin "thujone" is illegal to sell in the U.S.*

36" in height and spread. It has feathery, silver-gray foliage and inconspicuous flowers. This cultivar is reliably hardy to Zone 6, but with winter protection in a sheltered site it is worth trying in colder regions.

Problems & Pests

Rust, downy mildew and other fungi can cause problems for artemisias.

The genus name may honor any of the following: Artemisia, a botanist and medical researcher from 353 BC; Artemisia, the mourning widow of King Mausolus of Caria; or Artemis, goddess of the hunt and the moon in Greek mythology.

A. *ludoviciana* (above)
A. *ludoviciana* cultivar (below)

Arum
Arum

Also Called: Italian Arum

Height: 8–12" **Spread:** 8–12" **Flower color:** greenish white **Blooms:** early summer **Zones:** 5–9

THE FLASHY, SPOTTED, SPEAR-SHAPED LEAVES OF THIS ITALIAN woodland native appear in late fall along with even flashier stalks of plump, red-orange berries. The spring jack-in-the-pulpit-type spathes may appear pale and rather inconspicuous among all that goes on in a garden in spring, but those leaves and astonishing berries make up for its shy early display come November. Plant them where you can enjoy them best from fall through February, along with the hellebores, snowdrops and other winter show-offs.

Planting
Seeding: In a cold frame in fall

Planting out: Fall

Spacing: 8–12"

Growing

Arum grows well in **full sun** and
partial shade. The soil should be
fertile, humus rich and moist. Keep
in mind that this plant sprouts in
fall when deciduous trees are losing
their leaves and goes dormant just
as they leaf out in spring, so a location you consider shaded may be
appropriate if it is sunny at the right
time of the year. Divide it in summer
once flowering is finished. The pulp
should be removed from the seeds
before planting them. Be sure to
wear gloves when handling the plant
or its seeds because the sap and pulp
can irritate the skin and cause stom-
ach upset if ingested.

Tips

Arum is well suited to open woodland gardens and makes an excellent
ground covering companion for
shrubs. It is often included in pondside plantings where its attractive
green, fall-through-spring foliage
provides interest when most other
plants are dormant.

Recommended

A. italicum produces arrow-shaped
or sometimes spear-shaped, green-
or white-veined or marbled leaves.
It grows 8–12" tall with an equal
spread. Leaves emerge in fall and die
back in late spring. Greenish white
flower spathes are produced in early
summer, followed by colorful spikes
of bright reddish orange berries.
Subsp. 'Marmoratum' ('Pictum')
has distinctive white marbling on
the leaves.

Problems and Pests

Rarely suffers from problems.

A. italicum (all photos)

Aster

Aster

Height: 10"–5' **Spread:** 18–36" **Flower color:** red, white, blue, purple, pink; often with yellow centers **Blooms:** late summer to mid-fall **Zones:** 3–8

IF YOU CAN'T WAIT TO GET STARTED ON THE ASTER SEASON, THE white woodland aster, *Aster divaricatus*, blooms in summer and even better in shade. It can liven up your hosta display. Most people, though, think of asters as fall flowers. 'Purple Dome' is a terrific addition to the roster of New England asters, as it is not a tall, blowsy type like most but, as its name clearly advertises, a consistently well-disciplined dome, a nice change for the border and a must for control freaks.

Planting

Seeding: Not recommended

Planting out: Spring or fall

Spacing: 18–36"

Growing

Asters prefer **full sun** but tolerate partial shade. The soil should be **fertile, moist and well drained.** Pinch or shear these plants back in early summer to promote dense growth and reduce disease problems. Mulch in winter to protect plants from temperature fluctuations. Divide every two to three years to maintain vigor and control spread.

Tips

Asters can be used in the middle of borders and in cottage gardens. These plants can also be naturalized in wild gardens.

Recommended

A. divaricatus (white wood aster) is a spreading aster that tolerates shade. It grows 12–24" tall and spreads to 36". The flowers are tiny white stars with yellow centers. The stems tend to sprawl, but they can be controlled by pruning them by half in early to mid-June.

A. novae-angliae (Michaelmas daisy, New England aster) is an upright, spreading, clump-forming perennial that grows to 5' and spreads 24". It bears yellow-centered, purple flowers. **'Alma Potschke'** bears bright salmon pink or cherry red flowers. It grows 3–4' tall and spreads 24". The dwarf **'Purple Dome'** bears dark purple flowers. It grows 18–24"

A. novi-belgii (above & below)

tall and spreads 24–30". This cultivar resists mildew.

A. novi-belgii (Michaelmas daisy, New York aster) is a dense, upright, clump-forming perennial with purple flowers. It grows 3–4' tall and spreads 18–36". **'Alice Haslam'** is a dwarf plant with bright pink flowers. It grows 10–18" tall and spreads 18". **'Professor Kippenburg'** bears bright pink, double flowers on dwarf plants that grow 12–14" tall and spread about 18".

These old-fashioned flowers were once called starworts because of the many petals that radiate out from the center of the flowerhead.

A. novae-angliae (above), *A. novi-belgii* (forefront, below) & *A. novae-angliae* (background, below)

A. oblongifolius (aromatic aster) forms an mound of fragrant foliage. It grows 24" tall, spreads about 36" and bears blue flowers with yellow centers in mid- to late fall. **'Fanny'** bears loads of blue flowers on 3–4' plants that spread to 4'.

Problems & Pests

Powdery mildew, aster wilt, aster yellows, aphids, mites, slugs and nematodes can cause trouble.

What looks like a single flower of an aster, or of other daisy-like plants, is actually a cluster of many tiny flowers. Look closely at the center of the flowerhead and you will see all the individual florets

A. novae-angliae (above), *A. novi-belgii* (below)

Baptisia

Baptisia

Also Called: False Indigo

Height: 3–5' **Spread:** 2–4' **Flower color:** purple-blue, white
Blooms: late spring, early summer **Zones:** 3–9

THE RATTLE OF BAPTISIA SEEDS INSIDE THEIR SHELLS SPELLS THE end of the gardening season for some. The purple-black stems of wild white indigo, *B. alba*, create a striking contrast with the white, pea-like flowers—a truly beautiful plant. At 24–36" tall, they provide handsome foliage as a season-long backdrop for the mid-border. Great for dry gardens, baptisias are easier for most people to grow than lupines, and offer the same form and texture to a border design. Unsafe garden sex and deliberate breeding programs have resulted in an ever-widening array of new cultivars from which to choose each year.

Planting

Seeding: Sow indoors in early spring or direct sow in late summer; protect plants for the first winter

Planting out: Spring

Spacing: 24–36"

Growing

Baptisias prefer **full sun** but tolerate partial shade. Too much shade results in lank growth that causes the plants to split and fall. The soil should be of **average** or **poor fertility, sandy and well drained**. Baptisias are happy to remain in the same place for a long time and often resent being divided. Divide carefully in spring, only when you desire more plants.

Staking may be required if your plant is not getting enough sun. To prevent having to worry about staking or moving the plant, place it in the sun and give it lots of space to spread.

The hard seed coats may need to be penetrated before the seeds can germinate. Scratch the seeds between two pieces of sandpaper before planting them.

Tips

Baptisias can be used in an informal border or cottage-type garden. Use them in a natural planting, on a slope or in any well-drained, sunny spot. When first planted, baptisias may not look too impressive, but once established they are long-lived, attractive and dependable.

Recommended

B. alba is similar in size and form to *B. australis* (the more commonly grown baptisia). *B. alba* bears white flowers and is more tolerant of partial shade.

B. australis is an upright or somewhat spreading, clump-forming

B. australis 'Purple Smoke' (above), *B. australis* (below)

plant that bears spikes of purple-blue flowers.

B. **'Purple Smoke'** forms a dense mound of gray green foliage. It bears bi-colored purple and violet-blue flowers.

Problems & Pests

Minor problems with mildew, leaf spot and rust can occur.

Black-Eyed Susan
Rudbeckia

Also Called: Rudbeckia, Coneflower

Height: 18"–8' **Spread:** 12–36" **Flower color:** yellow or orange, with brown or green centers **Blooms:** mid-summer to fall **Zones:** 3–9

FROM AN AIR-CONDITIONED VANTAGE POINT OR A GARDEN CHAISE with cool drink in hand, watching goldfinches feast on the seeds of yellow coneflowers is a perfect activity in late summer when you've pretty much had it with gardening in the heat. Whether you plant black-eyed Susan, possibly the most common of all American wildflowers, or one of its taller, more elegant relatives, rudbeckias are cheerful and undemanding and make a bright statement when many other plants have gone on summer vacation. They enjoy the company of graceful ornamental grasses, Joe-Pye weed, sedums, globe thistle, penstemon 'Husker Red', dark-leaved shrubs such as DIABOLO ninebark (*Physocarpus*), BLACK BEAUTY elderberry (*Sambucus*) or purple smoke-bush (*Cotinus*), which also give the birds nice places to hide between snacks.

Planting

Seeding: Start seed in a cold frame or indoors in early spring; soil temperature should be about 61°–64° F

Planting out: Spring

Spacing: 12–36"

Growing

Black-eyed Susans grow well in **full sun** or **partial shade**. The soil should be of **average fertility and well drained**. Fairly heavy clay soils are tolerated, with several *Rudbeckia* species touted as 'claybusters.' Regular watering is best, but established plants are drought tolerant.

Pinch plants in June to encourage shorter, bushier growth. Divide in spring or fall every three to five years.

Tips

Use black-eyed Susans in wildflower or naturalistic gardens, borders and cottage-style gardens. They are best planted in masses and drifts.

Deadheading early in the flowering season keeps the plants flowering vigorously. Leave seedheads in place later in the season to add late-fall and winter interest and to provide food for birds.

Recommended

R. fulgida is an upright, spreading plant. It grows 18–36" tall and spreads 12–24". The orange-yellow flowers have brown centers. **Var.** *sullivantii* **'Goldsturm'** bears large, bright golden yellow flowers.

R. **'Goldquelle'** ('Gold Fountain') forms a large, open clump of

R. fulgida (above), *R. nitida* (below)

divided leaves, grows 36" tall and has bright yellow, double flowers.

R. **'Herbstsonne'** ('Autumn Sun') forms a large, upright clump bearing bright golden yellow flowers. It can grow 5–8' tall with a spread of 3–4'. Some support may be needed in a breezy location.

R. nitida is an upright, spreading plant 3–6' tall and 24–36" in spread. The yellow flowers have green centers. **'Autumn Glory'** has golden yellow flowers.

Problems & Pests

Rare problems with slugs, aphids, rust, smut and leaf spot are possible.

Bleeding Heart
Dicentra

Also Called: Dutchman's Breeches, Squirrel Corn
Height: 1–4' **Spread:** 12–36" **Flower color:** pink, white, red
Blooms: spring, summer **Zones:** 3–9

IN LATE SPRING THROUGH EARLY SUMMER, "BLEEDING" ROSE-COLORED "hearts" dangle enticingly from arching flower stems of the delightful Asian import, *Dicentra spectabilis*. Afterward, all goes dormant and disappears, so don't plant anything on top of its head! The charming flowers, suspended above gray-green leaves, have long been a cottage garden favorite. The native Dutchman's breeches (*Dicentra cucullaria*) was once used as a seductive love charm by Menomini Indian males, who believed that chewing the root and breathing the fragrance over one's "target audience" would cause the women to irresistibly follow the men anywhere. Another dainty native, dubbed squirrel corn (*D. canadensis*), has small, tubular, white flowers tipped with rose, resembling kernels of corn. The graceful, delicate flowers and ferny foliage of *D. eximia* bloom sporadically, spring through fall.

These delicate plants are perfect additions to a moist woodland garden. Plant them next to a shaded pond or stream.

Planting

Seeding: Start freshly ripened seed in cold frame; plants self-seed in the garden

Planting out: Spring

Spacing: 18–36"

Growing

Bleeding hearts prefer **light shade** but tolerate partial or full shade. The soil should be **humus rich, moist and well drained**. Though these plants prefer soil to remain evenly moist, they tolerate drought quite well when the weather isn't too hot. Very dry summer conditions cause the plants to die back. They will revive in fall or the following spring. It is most important for bleeding hearts to remain moist while blooming in order to prolong the flowering period. Regular watering will keep the flowers coming until mid-summer.

Bleeding heart rarely needs dividing.

Tips

Bleeding hearts can be naturalized in a woodland garden or grown in a border or rock garden. They make excellent early-season specimen plants and do well near a pond or stream.

All bleeding hearts contain toxic alkaloids, and some people develop allergic skin reactions from contact with these plants.

D. spectabilis (above & below)

D. eximia (above), *D. cucullaria* (center & below)

Recommended

D. 'Adrian Bloom' forms a compact clump of dark gray-green foliage. It grows about 12" tall and spreads about 18". Bright red flowers bloom in late spring and continue to appear intermittently all summer.

D. canadensis (squirrel corn) forms a small mound of feathery foliage. It grows about 12" tall with an equal spread. It bears white flowers in spring.

D. cucullaria (Dutchman's breeches) forms a small clump of delicate foliage. It grows about 8" tall and spreads up to 12". White flowers with yellow tips are borne in mid-spring.

D. eximia (fringed bleeding heart) forms a loose, mounded clump of lacy, fern-like foliage. It grows 15–24" tall and spreads about 18". The pink or white flowers are borne mostly in spring but may be produced sporadically over summer. Unless kept well watered, the plant will go dormant

during hot, dry weather. **'Alba'** bears white flowers.

D. **'Luxuriant'** is a low-growing hybrid with blue-green foliage and red-pink flowers. It grows about 12" tall and spreads about 18". Flowers appear in spring and early summer.

D. spectabilis (common bleeding heart, Japanese bleeding heart) forms a large, elegant mound up to 4' tall and about 18" in spread. It blooms in late spring and early summer. The inner petals are white and the outer petals are pink. This species is likely to die back in summer heat and prefers light-dappled shade. **'Alba'** has entirely white flowers. **'Goldheart'** has chartreuse foliage.

D. **'Stuart Boothman'** is a spreading perennial with blue-gray foliage. It grows about 12" tall, with an equal or greater spread. Dark pink flowers are produced over a long period from spring to mid-summer.

Problems & Pests

Slugs, downy mildew, *Verticillium* wilt, viruses, rust and fungal leaf spot can cause occasional problems.

The flowers of these plants are a popular addition to floral arrangements.

D. spectabilis 'Alba' (above), *D. eximia* 'Alba' (center)

D. eximia (below)

Blue Star Flower

Amsonia

Height: 24–36" **Spread:** 18"–4' **Flower color:** blue **Blooms:** late spring, early summer, mid-summer **Zones:** 3–9

AN ATTRACTIVE DEER-PROOF NATIVE, BLUE STAR FLOWER HAS graceful arching branches of linear, grass-like leaves. In late spring or early summer, it bears terminal clusters of pretty, pale blue, star-shaped flowers. In autumn, its most beautiful phase, foliage turns vibrant golden yellow and persists for almost a month. Fast becoming a favorite for designers, land-scape contractors and homeowners for its versatility and fine-textured appearance, it is reliable for many years. Blue star flower grows well in light shade, but in full sun, either as a specimen or en masse, this low-mainte-nance, disease-resistant plant offers textural contrast to the more common full-sun performers. Try growing blue star flower in combination with sweet-spire (*Itea virginica* LITTLE HENRY or *I. virginica* 'Henry's Garnet') for a stun-ning fall display.

A. hubrectii

Planting

Seeding: Start in a cold frame in fall

Planting: Spring or fall

Spacing: 18–24"

Growing

Blue star flowers grow well in **full sun, partial shade and light shade**. The soil should be **average to fertile, moist and well drained**. Plants in light or partial shade locations are more drought tolerant than those grown in full sun. Divide in early fall or spring. Shear plants back after blooming is finished to deadhead, keep the plants looking tidy and encourage bushy growth. Wash your hands after handling these plants as the sap irritates some people's skin.

Tips

Blue star flower makes an excellent addition to a mixed or herbaceous border. Grow it along the edge of a woodland or near a water feature, or allow it to naturalize in a meadow garden because it is quite prone to prolific self-seeding.

Recommended

A. hubrectii (Arkansas blue star) is a clump-forming plant with narrow green leaves that turn bright yellow in fall. It grows 24–36" tall and spreads $2^{1}/_{2}$–4'. Clusters of blue flowers are produced in late spring or early summer. (Zones 5–9)

A. tabernaemontana (common blue star) forms a shrubby clump of dark green leaves. It grows 24–36" tall and spreads 18–30". Clusters of pale blue flowers are borne from late spring to mid-summer. (Zones 3–9)

Problems and Pests

Rare problems with rust are possible.

Boltonia
Boltonia

Also Called: False Aster

Height: 3–6' **Spread:** 4' or more **Flower color:** white, mauve or pink, with yellow centers **Blooms:** late summer and fall **Zones:** 4–9

I HAVE FOND MEMORIES OF A GREAT SHOW I HAD FOR A FEW YEARS —billowing clouds of tiny daisies on tall blue-green stems way above my head, blooming from July through September in three huge mid-border clumps of boltonia, surrounded by sedum, Stokes' aster, iris, daylilies and coral bells with morning glories climbing boltonia's tall stems. Alas, by fall they fell apart, and boltonia departed completely after a few summers. Wish I knew why. Other stands of this plant have been reappearing faithfully but not as exuberantly for decades, so I must remember to replant and, perhaps, recreate that advertisement for why everyone should try boltonia in their garden at least once.

Planting

Seeding: Start seed in cold frame in fall

Planting out: Spring or fall

Spacing: 36"

Boltonia is native to the eastern and central United States.

Growing

Boltonia grows best in **full sun** but tolerates partial shade. It prefers soil that is **fertile, humus rich, moist and well drained** but adapts to less fertile soils and tolerates some drought. Divide in fall or early spring when the clump becomes overgrown or begins to die out in the middle.

Staking may be required, particularly in partially shaded areas. Use twiggy branches or a peony hoop to provide support. If installed while the plant is young, the stakes will be hidden by the growing branches. Alternatively, cut the plant back by one-third in June to reduce the height and prevent flopping.

Tips

This large plant can be used in the middle or at the back of a mixed border, in a naturalized or cottage-style garden or near a pond or other water feature.

B. asteroides (above & below)

Boltonia is a fine fresh-cut flower for arrangements, a great butterfly nectar source and a hummingbird magnet.

A good alternative to Michaelmas daisies, boltonia is less susceptible to powdery mildew.

Recommended

B. asteroides is a large, upright perennial with narrow, grayish green leaves. It bears many white or slightly purple, daisy-like flowers with yellow centers. **'Pink Beauty'** has a looser habit and bears pale pink flowers. It is less vigorous than **'Snowbank,'** which has a denser, more compact habit and bears more plentiful white flowers than the species.

Problems & Pests

Boltonia has rare problems with rust, leaf spot and powdery mildew.

Brunnera
Brunnera

Also Called: Siberian Bugloss

Height: 12–18" **Spread:** 18–24" **Flower color:** blue **Blooms:** spring
Zones: 3–8

TUCK THIS LUSH SPREADER, VALUED BOTH FOR FOLIAGE AND
flowers, among your ferns and hostas. Tiny, deep blue, forget-me-not look-
alike flowers bloom from early spring through Memorial Day. The leaves on
some of the newer cultivars are quite sensational. One that I grow, 'Jack Frost,'
is an absolute gem and easily cultivated. I've seen it planted in great sweeps,
where it virtually glows. 'Looking Glass' would shimmer in the woodland,
too. Well adapted to moist areas edging wetlands, Siberian bugloss offers a
worthy solution for low, damp, shady spots. Mulch to keep roots cool in hot
weather to prevent premature heat-induced dormancy.

Planting

Seeding: Start seed in cold frame in early fall or indoors in early spring

Planting out: Spring

Spacing: 12–18"

Growing

Brunnera prefers **light shade** but tolerates morning sun with consistent moisture. The soil should be of **average fertility, humus rich, moist and well drained**. The species and its cultivars do not tolerate drought.

Cut back faded foliage mid-season to produce a flush of new growth. Divide in spring when the center of the clump begins to die out.

Tips

Brunnera makes a great addition to a woodland or shaded garden. Its low, bushy habit makes it useful as a groundcover or as an addition to a shaded border.

Recommended

B. macrophylla forms a mound of soft, heart-shaped leaves and produces loose clusters of blue flowers all spring. '**Jack Frost**' has platinum leaves with dark green veins. '**Langtrees**' has pewter-colored marks on the leaves. Grow variegated plants in light or full shade to avoid scorched leaves. '**Looking Glass**' has silver foliage with a silken sheen and delicate green veining.

This reliable plant rarely suffers from any problems.

B. macrophylla (above), *B. macrophylla* 'Jack Frost' (below)

Bugbane

Actaea (Cimicifuga)

Also Called: Black Cohosh, Black Snakeroot, Fairy Candles
Height: 2–7' **Spread:** 24" **Flower color:** cream, white, pink
Blooms: late summer to late fall **Zones:** 3–8

THIS PLANT'S MANY COMMON NAMES HAVE AN INTERESTING range of meanings. "Bugbane" refers to the plant's insect repellant qualities. "Black cohosh" is traceable to Native Americans, who used this plant for treating "female complaints." (Imagine! Tribal males must have left their socks on the floor and never asked for directions either!) *Cimicifuga*, as it was once known, is thought to have numerous other medicinal applications, and is currently under scrutiny as a preventative and treatment for breast cancer. "Fairy candles," naturally, are used by wee folk to illuminate the garden during nightly frolics. As if all the names weren't confusing enough, there is another woodland plant under the genus *Actaea*, known commonly as doll's eyes or baneberry. Rest assured, taxonomists will undoubtedly change something again soon.

Planting

Seeding: Start fresh seed in cold frame in fall; should germinate the following spring

Planting out: Spring

Spacing: 18–24"

Growing

Bugbanes grow best in **partial** or **light shade.** The soil should be **fertile, humus rich and moist.** Plants may require support from a peony hoop. The roots resent being disturbed, so the plants should not be divided. Bugbanes spread by rhizomes, and small pieces of root can be unearthed carefully and replanted if more plants are desired.

Tips

These plants make attractive additions to an open woodland garden, shaded border or pondside planting. They don't compete well with tree roots or other vigorous-rooted plants. They are worth growing close to the house because the late-season flowers are wonderfully fragrant.

Bugbane adds desirable architectural interest to partially shaded gardens with its towering stems and slender, white, mid-summer flower wands. Plant ferns, coral bells and hostas at its feet.

Recommended

A. racemosa is a clump-forming, native perennial 4–7' tall and about 24" wide. Long-stemmed spikes of fragrant white flowers are borne from late summer into fall. There are purple-leaved cultivars available

A. racemosa (above), *A. simplex* cultivar (below)

for both of these species, but they tend to revert to green, especially when grown in too deep a shade.

A. simplex is also a clump-forming perennial. It grows 3–4' tall and spreads about 24". The scented, bottlebrush-like spikes of flowers are borne in fall.

Problems & Pests

Occasional problems are possible with rust and with insect damage to the leaves.

Campanula
Campanula

Also Called: Bellflower, Harebell

Height: 4"–6' **Spread:** 12–36" **Flower color:** blue, white, purple
Blooms: summer, early fall **Zones:** 3–7

BELLFLOWERS ARE A LOVELY LOT, AND AN ENORMOUS GENUS, ONLY a handful of which have reached the garden, some cultivated since medieval times. Over 300 species of Campanula grow throughout the Northern Hemisphere, in habitats ranging from high, rocky crags to boggy meadows. Sadly, some of the finest ones (*C. lactiflora* and *C. latifolia,* for instance) grow to their full magnificence in the British Isles or, perhaps, our West Coast, but conduct themselves disappointingly here. And some of the star performers here (*C. punctata, C. parryi, C. takesimana* and *C. rapunculoides*) are so invasive, they will ring their bells while racing you across the garden, and exhaust you in the end. When I have a lust for tall, bell-shaped beauties, I stick to clematis, foxgloves and lady bells (*Adenophora.*) The species described below are a well-mannered few.

Planting

Seeding: Not recommended because germination can be erratic; if desired, direct sow in spring or fall

Planting out: Spring or fall

Spacing: 12–36"

Growing

Campanulas grow well in **full sun, partial shade** or **light shade**. The soil should be of **average to high fertility and well drained**. These plants appreciate a mulch to keep their roots cool and moist in summer and protected in winter, particularly if snow cover is inconsistent. It is important to divide campanulas every few years in early spring or late summer to keep plants vigorous and to prevent them from becoming invasive.

Deadhead to prolong blooming. For the upright campanulas, consider cutting back sections of the plant one at a time. Doing so allows for new buds to form on the pruned section, which will bloom as the uncut sections finish blooming.

Campanulas can be propagated by basal, new-growth or rhizome cuttings.

Tips

Upright and mounding campanulas can be used in borders and cottage gardens. Low, spreading and trailing campanulas can be used in rock gardens and on rock walls. You can also edge beds with the low-growing varieties.

Recommended

C. **'Birch Hybrid'** is low growing with a spreading habit, reaching 4–6" in height and 12" in spread. It bears light blue to mauve flowers in summer.

C. glomerata (clustered bellflower) forms a clump of upright stems. It grows 12–24" tall, with an equal or

C. glomerata 'Superba'

greater spread. Clusters of purple, blue or white flowers are borne over most of the summer. **'Joan Elliott'** bears large, rich violet flowers. **'Superba'** has dark purple flowers.

C. portenschlagiana (Dalmatian bellflower) is a low, spreading, mounding perennial 6" tall and 20–24" in spread. It bears light or deep blue to violet blue flowers from mid- to late summer.

C. poscharskyana (Serbian bellflower) is a trailing perennial that likes to wend its way through other plants. It grows 6–12" tall, spreads 24–36" and bears light violet blue flowers in summer and early fall.

Problems & Pests

Minor problems with vine weevils, spider mites, aphids, powdery mildew, slugs, rust and fungal leaf spot are possible.

Catmint

Nepeta

Height: 10–36" **Spread:** 18–36" **Flower color:** blue, purple, white, pink **Blooms:** spring, summer, occasionally again in fall **Zones:** 3–8

CATS AND HUMANS, ESPECIALLY HERBALISTS, LOVE CATMINT AND deer don't. That might be enough of a recommendation, but the gray-green, airy foliage of this plant adds informality, color and textural respite from any stiff and strident planting. All cultivars should be sheared after blooming to promote bushy growth and keep the plant from flopping open in the middle. Of course, when a cat decides to plunk himself down for a nap in your catmint, what's a gardener to do? It's tempting to join right in. Catmints are great bee plants, though, so proceed with caution. I like to plant them so they can hang over the wall of a raised bed. That way, kitty and I are on a level playing field.

Planting

Seeding: Most popular hybrids and cultivars are sterile and cannot be grown from seed

Planting out: Spring
Spacing: 18–24"

Growing

Catmint grows well in **full sun** or **partial shade**. The soil should be of **average fertility and well drained**. The growth tends to be floppy in too rich a soil.

Pinch tips in June to delay flowering and make the plants more compact. Once plants are almost finished blooming, you may cut them back by one-third to one-half to encourage new growth and more blooms in late summer or fall. Divide in spring or fall when the plants begin to look overgrown and dense.

Tips

Catmint can be used to edge borders and pathways. They work well in herb gardens and with roses in cottage gardens. Taller varieties make lovely additions to perennial beds, and dwarf types can be used in rock gardens.

Think twice before growing *N. cataria* (catnip), as cats are heavily attracted to this plant and may wander into your garden to enjoy it. Cats do like the other catmints as well, but not to quite the same extent.

Recommended

N. x *faassenii* forms a clump of upright and spreading stems. It grows 18–36" tall, with an equal spread, and bears spikes of blue or purple flowers. This hybrid and its cultivars are sterile and cannot be grown from seed. **'Alba'** bears white flowers. **'Dropmore'** has gray-green foliage and light purple flowers. **'Snowflake'** is low growing, compact and spreading, with white flowers. It grows 12–24" tall and spreads about 18". **'Walker's Low'** has gray-green foliage and bears lavender blue flowers. It grows about 10" tall.

N. sibirica **'Souvenir d'André Chaudron'** is a low, mound-forming perennial. It grows 12–18" tall with an equal spread. It bears violet blue flowers in spring and early summer.

N. **'Six Hills Giant'** is a large, vigorous plant about 36" tall and about 24" tall. It bears large, showy spikes of deep lavender blue flowers.

Problems & Pests

These plants are pest free, except for an occasional bout of leaf spot or a visit from a neighborhood cat.

N. x faassenii 'Dropmore'

Chrysanthemum

Chrysanthemum

Also Called: Fall Garden Mum, Hybrid Garden Mum

Height: 12–36" **Spread:** 2–4' or more **Flower color:** orange, yellow, pink, red, purple **Blooms:** late summer and fall **Zones:** 5–9

FALL MUMS WERE A GARDEN STAPLE WHEN I LIVED IN MARYLAND, coming back year after year. Here in Pennsylvania, not that much further north, I have to consider them annuals, except for the Siberian daisy, *C. zawadskii*, now found under the alias 'Clara Curtis,' which has lingered for me and proliferated for almost 20 years. It's a delightfully fragrant late bloomer (sometimes in November.) The bees get their last licks in when it's in flower. Another chrysanthemum that deserves more hype is the ground-cover mum, *C. weyrichii*. It, too, is a true perennial, drought tolerant and carefree, which forms a beautiful year-round carpet, blooming in late summer through autumn under red-berried heavenly bamboo (*Nandina*), sweetspire (*Itea*), blue star flower (*Amsonia*) and colored-up maples (*Acer*).

Planting

Seeding: Not recommended
Planting out: Spring, summer or fall
Spacing: 18–24"

Growing

Chrysanthemums grow best in **full sun**. The soil should be **fertile, moist and well drained**. The earlier in the season you can plant chrysanthemums, the better. Early planting improves their chances of surviving the winter.

Pinch plants back in early summer to encourage bushy growth. In late fall or early winter, you can deadhead the spent blooms, but leave the stems intact to protect the crowns of the plants. Divide in spring or fall every two years to keep plants vigorous and to prevent them from thinning out in the center.

Tips

These plants provide a blaze of color in the late-season garden, often flowering until the first hard frost. Dot or group them in borders or use them as specimen plants near the house or in large planters. Some gardeners purchase chrysanthemums as flowering plants in late summer and put them in the spots where summer annuals have faded.

Recommended

C. **'Mei-Kyo'** is a vigorous grower that blooms in mid- to late October. The flowers are a deep pink. Be sure to pinch back the stems to prevent the need to stake. Plants grow to 24–30" tall and spread as far as you will let them.

C. hybrids (above & below)

C. **'Morden'** series was developed in Canada and is reliably hardy to Zone 4. Plants come in a wide variety of colors and grow about 24" tall.

C. **'My Favorite'** series is a new introduction and is heralded as a truly perennial mum and a prolific flower producer. The plants grow to about 24" tall and spread about 4'.

C. **'Prophet'** series is popular and commonly available. Plants grow about 24" tall and spread about 24–36". Flowers come in all colors.

C. **Rubellum Group** cultivars grow to 24–30" in height and spread. **'Clara Curtis'** has soft pink flowers; **'Mary Stoker'** is a vigorous grower with butter yellow flowers.

C. weyrichii (*Dendranthema weyrichii*; dwarf chrysanthemum) forms a low, creeping mat of foliage. It grows up to 12" tall and spreads 18–24". The daisy-like autumn flowers

C. Rubellum Group 'Clara Curtis' (above), C. hybrids (below)

have white or pink petals with yellow centers.

Problems & Pests

Aphids can be a true menace to these plants. Insecticidal soap can be used to treat the problem, but it should be washed off within an hour because it discolors the foliage when the sun hits it. Also watch for spider mites, whiteflies, leaf miners, leaf spot, powdery mildew, rust, aster yellows, blight, borers and rot, though these problems are not as common.

Although the name Chrysanthemum *comes from the Greek for "golden flower," these plants come in a wide range of bright colors.*

C. Prophet series 'Raquel' (above)
C. Prophet series 'Stacy' (below)

Columbine
Aquilegia

Height: 9–36" **Spread:** 12–18" **Flower color:** red, yellow, pink, purple, blue, white; color of spurs often differs from that of petals
Blooms: spring, summer **Zones:** 2–9

A DENIZEN OF ROCKY CLIFFS AND OUTCROPPINGS, COLUMBINES are short-lived perennials that seed freely, establishing themselves in unexpected, and often charming, locations. If you wish to keep a particular one, preserve it carefully through frequent division or root cuttings. The dainty, suspended flowers of columbines attract butterflies and hummingbirds. The common European columbine, from which many were bred, is blue, but our native one, *A. canadensis*, is a snazzy red-and-yellow duet and hard to top for its elegant form and graceful garden presence. Plant these flowers "up close and personal" where you can admire their little nodding flowers. Try planting the wild columbine with Indian Pink (*Spigelia marilandica*) for a color match with a textural twist.

Planting

Seeding: Direct sow in fall or spring

Planting out: Spring

Spacing: 18"

Growing

Columbines grow well in **partial shade**. They prefer soil that is **fertile, moist and well drained**, but they adapt well to most soil conditions. Division is not required but can be done to propagate desirable plants. The divided plants may take a while to recover because they dislike having their roots disturbed.

A. flabellata (above), *A. canadensis* (below)

Tips

Use columbines in rock gardens, formal or casual borders and naturalized or woodland gardens.

Columbines self-seed but are in no way invasive. Each year a few new seedlings may turn up near the parent plant and can be transplanted. If you have a variety of columbines planted near each other, you may even wind up with a new hybrid, because these plants readily crossbreed. A wide variety of flower colors is the most likely and interesting result. Because many columbines grown in gardens are hybrids, the new seedlings may not be identical to the parent. They may revert to one of the original parent species.

Recommended

A. **'Biedermeier'** is a compact hybrid, forming a low mound 9–12" tall with white, purple or pink flowers.

A. chrysantha (above), *A. vulgaris* 'Nora Barlow' (below)

*A. **canadensis*** (wild columbine, Canada columbine) is native to Pennsylvania and most of eastern North America and is common in woodlands and fields. It grows up to 24" tall, spreads about 12" and bears yellow flowers with red spurs. **'Corbett'** grows about 18" tall and bears light yellow flowers.

*A. **chrysantha*** (golden columbine, yellow columbine) in an upright plant that grows up to 36" tall and spreads about 24". It bears yellow flowers from late spring to late summer, blooming best when regularly deadheaded. **'Yellow Queen'** bears larger bright yellow flowers with long spurs.

*A. **flabellata*** (fan columbine, Japanese fan columbine) is a dwarf plant with blue-green foliage that is resistant to leaf miners and grows 4–12" tall and spreading 4–6". Flowers were originally only available in blue, but are now available in a range of pastel shades of pink and white.

A. **'McKana Giants'** (McKana Hybrids) are vigorous plants, growing up to 36" tall and bearing flowers in yellow, pink, red, purple, mauve and white.

Aquilegia *means "drawing water" in Latin, and refers to nectar that collects in the spurs. "Columbine" probably comes from the Renaissance Italian* commedia dell'arte, *farcical and comedic plays where a stock character was the wildly and comically dressed Columbina, the mistress of Harlequin.*

A. vulgaris (European columbine, common columbine) grows about 36" tall and spreads 18". This species has been used to develop many hybrids and cultivars with flowers in a wide variety of colors. **'Black Barlow'** has fully double, spurless, near-black flowers that resemble dahlias. **'Nora Barlow'** is a popular cultivar with double flowers in white, pink and green-tinged red.

Problems & Pests

Mildew and rust can be troublesome during dry weather. Other problems include leaf miners, fungal leaf spot, aphids and caterpillars. If the foliage becomes ragged from leaf miner attack, simply cut it back and burn it. Do not put infected leaves into the compost pile, or the leaf miner larvae will overwinter and emerge as adult flies to infect next year's columbines.

Some Native American males would coat their hands with powdered columbine seeds before holding hands with desired companions. Washing hands together afterward ensured the couple a long, happy life together.

A. 'McKana Giants' (above & below)

Coral Bells

Heuchera

Also Called: Heuchera, Alumroot

Height: 6–36" **Spread:** 6–18" **Flower color:** red, pink, white, purple; plant also grown for foliage interest **Blooms:** spring, summer **Zones:** 3–9

FLOWERS OF HEUCHERA (CORRECTLY PRONOUNCED HOY-KER-UH (here's your chance to be one of the few living gardeners to know this) are light and airy, white, pink, coral, lilac or red variations. They're showy in some cases, inconsequential in others, compared to the lovely leaves. Some coral bells do better in shade; others prefer some sun. I grow most of mine in full sun, including my current favourite, deep purple—almost black—shiny 'Obsidian'. I'll admit I went a bit coral bell crazy last year, but who can resist plants dubbed 'Peach Flambé,' 'Caramel,' 'Marmelade,' 'Crème Brûlée,' 'Plum Pudding,' 'Mint Frost,' 'Key Lime Pie,' 'Champagne Bubbles' and —heaven help us—'Chocolate Ruffles'? (Thank goodness that isn't 'Truffles' or I'd be chomping my way across the garden right now.)

Planting

Seeding: Species, but not cultivars, may be started from seed in a cold frame

Planting out: Spring

Spacing: 12–18"

Growing

Coral bells grow best in **light to partial shade**. Many do well in **full sun**, although foliage colors can bleach out for others. Plants may become leggy in full shade. Check the label of your purchased cultivar for the recommended siting. The soil should be **rich, moist and well drained**. Good air circulation is essential.

Deadhead to prolong the bloom. If stems become woody, if the plant goes bald at the center or if flowering diminishes, divide and replant, preferably in spring, but fall is okay, too. Replant with the crown at or just above soil level.

Tips

Use coral bells as edging plants, in clusters in woodland gardens or as groundcovers in low-traffic areas. Combine foliage of different types for an interesting display. Coral bells also make excellent container plants.

Because of their shallow root systems, winter freezing and thawing tend to heave coral bells out of the soil. Mulching helps maintain an even soil temperature. Most crucial, though, never cover their crowns with soil or mulch, (You'd suffer, too, if someone buried your face.)

H. sanguinea 'Monet' (above)
H. 'Chocolate Ruffles' (below)

Recommended

H. americana is native to the central and eastern U.S. This mound-forming plant grows about 18" tall and 12" wide. Its heart-shaped foliage is marbled and bronze-veined when it is young and matures to deep green. Cultivars have been developed for their attractive and variable foliage. **'Dale's Strain'** has silver-mottled, blue-green leaves with reddish veins. Flowers are white.

***H.* Bressingham Hybrids** are compact hybrids that can be started from seed. The flowers are in shades of pink or red. Plants grow about 24" tall, with an equal spread.

***H.* 'Caramel'** has apricot-colored foliage and pink flowers. It grows about 18" tall, with an equal spread.

***H.* 'Chocolate Ruffles'** has ruffled, glossy brown foliage with purple undersides that give the leaves a bronzed appearance. It grows 10–30" tall and 18–24" wide.

H. 'Bressingham Bronze' (above), *H.* hybrid (below)

H. **'Coral Cloud'** forms a clump of glossy, crinkled leaves and bears pinkish red flowers. It grows about 30" tall and spreads about 18".

H. **'Firefly'** develops a clump of dark green leaves with attractive, fragrant, bright pinkish red flowers. It grows 13–30" tall and spreads 12–18".

H. **'June Bride'** is a mound-forming plant that grows 12–18" tall, with an equal spread. It bears large, white flowers.

H. **'Lime Rickey'** forms a low mass of chartreuse leaves. It grows about 8" tall and spreads about 18".

H. **'Marmalade'** has foliage that emerges red and matures to orange-yellow. It grows about 9" tall and spreads to about 18".

H. micrantha is a mounding, clump-forming plant that grows 18–24" tall with an equal spread. The foliage is

H. micrantha var. *diversifolia* 'Palace Purple' (above & below)

H. 'Marmalade' (above), H. 'Firefly' (below)

gray-green and the flowers are white. The species is not common in gardens but many of its cultivars and hybrids are. **Var.** *diversifolia* 'Palace Purple' is a well-known cultivar and was one of the first heucheras grown for its interesting foliage. This compact plant grows 18–20" tall and has deep purple foliage and white blooms. It can be started from seed, but only some of the seedlings will be true to type.

H. 'Montrose Ruby' has bronzy purple foliage with bright red undersides. It grows about 12" tall and spreads 18–24".

H. 'Northern Fire' has red flowers and leaves mottled with silver. It grows about 18" tall, with an equal spread.

H. 'Obsidian' has lustrous, dark purple, nearly black, foliage. It grows about 24" tall, with a spread of 12".

H. 'Pewter Moon' has light pink flowers and silvery leaves with bronzy purple veins and undersides.

H. 'Pewter Veil' has silvery purple leaves with dark gray veins. Its flowers are white flushed with pink. It grows 8–24" tall and spreads up to 24".

H. 'Raspberry Regal' is a mound-forming plant that grows up to 4' tall. The foliage is strongly marbled and the flowers are bright red.

H. sanguinea is the hardiest species. It forms a low-growing mat of foliage and reaches 6–18" in height and spread. The dark green foliage is marbled with silver. Red, pink or white flowers are borne in summer. 'Frosty' has red flowers and silver-variegated foliage. **'Monet'** has green leaves mottled with creamy white. **'White Cloud'** has silver-mottled leaves and bears white flowers in late spring.

H. **'Velvet Night'** has dark purple leaves with a metallic sheen. Flowers are creamy white. It grows 8–24" tall and spreads 18–24".

H. villosa is a mound-forming plant that grows about 18" tall and 12–18" wide. It bears white or pink flowers. **'Autumn Bride'** has light green, slightly fuzzy leaves. **'Purpurea'** has bronzy purple leaves.

H. sanguinea (above)
H. americana 'Plum Pudding' (below)

Problems & Pests

Healthy heucheras have very few problems. In stressed situations, they can be afflicted with foliar nematodes, powdery mildew, rust or leaf spot.

Alumroot has many medicinal applications and, like alum (aluminum sulphate), can be used as a mordant to fix colors when dyeing fabric.

Coreopsis
Coreopsis

Also Called: Tickseed

Height: 18–36" **Spread:** 12–18" **Flower color:** yellow, pink or orange
Blooms: early to late summer **Zones:** 3–9

AN EASY-TO-GROW SUNFLOWER RELATIVE, COREOPSIS CONTRIBUTES cheerful color throughout the gardening season. Since it can endure intense heat and tolerate soil of low fertility, it succeeds even on roof gardens in urban environments. Coreopsis provides an abundant source of cut flowers, and will bloom longer if deadheaded. The profusion of blossoms can make deadheading agonizing, though. Simply shear the whole plant back after the first flush of bloom declines, and the plant will spring right back. Because of its long bloom time and its exuberance, it is useful for filling bare spots in the border, pairing particularly well with blues and purples. Its character contrasts effectively with the bold foliage of many shrubs and other perennials. Coreopsis does nicely in containers and hanging baskets, too.

Planting

Seeding: Direct sow in spring. Seeds may be sown indoors in winter, but soil must be kept fairly cool, at 55°–61° F, for seeds to germinate.

Planting out: Bareroot plants are best planted in spring. Potted plants can go in the ground anytime from late spring until a month before a hard, killing frost.

Spacing: 18–24"

Growing

Grow coreopsis in **full sun**. The soil should be **average, sandy, light and well drained**. Plants can develop crown rot in moist, cool locations with heavy soil. Overly fertile soil causes long, floppy growth.

To deadhead use scissors to snip out tall stems, and shear plants by one-half in late spring for more compact growth. Frequent division may be required to keep plants vigorous and prolong their lives.

Tips

Coreopsis is a versatile plant, useful in formal and informal borders and in meadow plantings or cottage gardens. It looks best in groups. *C. grandiflora* self-seeds readily but is easy to remove from areas where it isn't wanted.

Mass plant coreopsis to fill in a dry, exposed bank where nothing else will grow, and you will enjoy the bright, sunny flowers all summer long.

C. grandiflora 'Early Sunrise' (above), *C. rosea* (below)

Recommended

C. grandiflora (large-flowered coreopsis, tickseed) forms a clump of foliage and bears bright golden yellow flowers over a long period in mid- and late summer. It grows 18–36" tall and spreads 12–18". This species and its cultivars are often grown as annuals because bearing so many flowers leaves them with little energy for surviving the winter. **'Early Sunrise'** is a compact plant that grows 18–24" tall. It bears double yellow flowers and can be started from seed. **'Mayfield Giant'** has large, intense yellow flowers on plants up to 36" tall.

C. rosea (pink tickseed) is an unusual species with pink flowers. It grows 24" tall and 12" wide. This species is more shade and water tolerant than the other species but is not as vigorous.

C. verticillata 'Moonbeam' (above), *C. grandiflora* cultivar with *C. verticillata* 'Moonbeam' (above)

C. verticillata (thread-leaf coreopsis) is a mound-forming plant with attractive, finely divided foliage. It grows 24–32" tall and spreads 18". **'Golden Showers'** has large, golden yellow flowers and ferny foliage. **'Moonbeam'** forms a compact mound of delicate foliage. The flowers are a light, creamy yellow. **'Zagreb'** is an upright plant that produces golden blooms over a long period of time.

Problems & Pests

Occasional problems with slugs, bacterial spot, gray mold, aster yellows, powdery mildew, downy mildew, crown rot and fungal spot are possible.

C. verticillata (above)
C. grandiflora 'Early Sunrise' (below)

Corydalis

Corydalis

Height: 8–18" **Spread:** 8–12" or more **Flower color:** yellow, cream, blue **Blooms:** spring, summer **Zones:** 5–7

POKING AROUND ONE OF THE HUGE GREENHOUSES IN THE Pennsylvania Dutch country, I was besotted by a mesmerizing fragrance wafting about. I pursued it till I located the source. Was it lilacs? Turned out it was *Corydalis* 'Silver Spectre.' I had no idea these ferny little ground huggers were so aromatic! I guess you'd have to be a mouse or slithering creature to ever notice it in the garden. I recommend planting corydalis on a wall where your nose can take full advantage of its fleeting but fabulous scent. Unfortunately, corydalis takes a vacation when the weather heats up. (Sigh.)

Planting

Seeding: Direct sow fresh seed in early fall; germination can be erratic

Planting out: Spring

Spacing: 12"

Growing

Corydalis plants grow well in **light** or **partial shade** with **morning sun**. The soil should be of **average to rich fertility, humus rich and well drained**.

Plants will die back in the hottest part of summer. Trim the faded foliage, and new leaves will sprout as the weather cools in late summer and fall. These plants self-seed and can be propagated by transplanting the tiny seedlings. Division can be done in spring or early summer, but corydalis resent having their roots disturbed.

Tips

Corydalis are admired for their delicate flowers and attractive, ferny foliage. Use them in woodland or rock gardens, in borders, on rock walls and along paths. Let them naturalize in unused or underused areas.

Recommended

C. flexuosa (blue corydalis) is an erect plant with fragrant, blue, spring flowers. It grows 12" tall and spreads 8" or more. Keep this plant well watered during hot weather. '**Blackberry Wine**' has blue-green foliage and fragrant, purple-blue flowers.

C. leucanthema '**Silver Spectre**' is a more heat-tolerant selection of corydalis. The green foliage has a silvery cast near the center of each leaf. The fragrant, pale mauve flowers have dark purple tips. It grows about 12" tall and about 36" wide.

C. lutea (yellow corydalis) is a mound-forming perennial that bears yellow flowers from late spring often to early fall. It grows 12–18"

C. lutea (above & below)

tall and spreads 12" or more. This one is the hardiest species; it is also the most vigorous and can become invasive. Most gardeners don't mind the ferny-leaved plants turning up here and there.

C. ochroleuca (white corydalis) bears cream to white flowers in late spring and summer and is very similar to *C. lutea* in habit. It grows about 12" tall, with an equal spread.

Problems & Pests

Rare problems with downy mildew and rust are possible.

Culver's Root
Veronicastrum

Also Called: Bowman's Root, Veronicastrum

Height: 4–6' **Spread:** 18–36" **Flower color:** white, pink, purple-blue
Blooms: mid-summer to fall **Zones:** 3–8

CULVER'S ROOT IS A STATELY PLANT THAT ADDS A DISTINCTIVE,
lofty, vertical accent to a composition of perennials or prairie plants. At the
height of summer, whorls of deep green leaves atop branching stems reach as
high as seven feet on occasion—a giant candelabrum. This plant is native to
meadows and moist thickets, and is easy to grow in adequately damp soil. It
also supplies good cut flowers for arrangements.

Planting

Seeding: Start in a cold frame in fall
Planting out: Spring
Spacing: 18–24"

Growing

Culver's root grows well in **full sun**
and **partial shade.** The soil should be
**average to fertile, humus-rich and
moist.** Divide plants in spring or fall
when clumps appear to be thinning in
the middle or when plants appear to
be growing less vigorously.

Tips

These large, vigorous plants add
height to the back and middle of a
mixed or herbaceous border. They
also make lovely additions to open
woodland gardens, meadow plantings, cottage-style gardens and in
the moist soil near a water feature.
Their late season blooming in pastel
shades is a welcome change from
the yellows, oranges and golds that
tend to dominate at this time of year.

Recommended

V. virginicum is an upright perennial that grows 4–6' tall with a
spread of 18–36". Spikes of fuzzy-
looking white, pink or purple-blue
flowers are produced from mid-
summer to early fall. **'Album'** (var.
album) bears white flowers. **Var.
*incarnatum*** ('Roseum') bears pink
flowers. **Var. *sibiricum*** is a super-
hardy variety with long, lavender
blue flower spikes.

Problems & Pests

Problems with powdery or downy
mildew and leaf spot can occur.

V. virginicum 'Album' (above)
V. virginicum var. *incarnatum* (below)

Cyclamen

Cyclamen

Height: 2–8" **Spread:** indefinite **Flower color:** white, pink, red
Blooms: late winter to early spring **Zones:** 5–9

CYCLAMEN HAVE DELIGHTFUL, VIVID, LITTLE FLOWERS THAT LOOK
like a field of tiny butterflies when in bloom, and at an unlikely time, too.
Pointed buds arise in fall to late winter, unfurling into splendid stands of
pink, white and magenta. Later the beautifully marked leaves arrive, some
dark green marbled silver with varied patterns, others like frosted pewter,
which add to the flowers' impact. Many hybrids have fragrant flowers.
Cyclamen thrives beneath camellias, rhododendrons and ferns and under
trees. All cyclamen, including the non-hardy florist's kind, grow best in rich,
extremely fast-draining soil. Planting them under trees allows the tree roots
to absorb excess moisture.

Planting

Seeding: Soak seeds in water for at least 12 hours before planting in a cold frame in fall. Keep seeds in the dark until they germinate.

Planting out: The best time to plant or transplant tubers is during cyclamen's dormant season, or with the autumn or winter flowering varieties, after they have finished flowering. Planting depth is not too critical since, like daffodils, they develop contractile roots and will pull themselves down to the proper depth, depending on local conditions. Planting them 2–3" deep is a good rule of thumb.

Spacing: 6"

Growing

These tuberous perennials grow best in **light** or **partial shade**. The soil should be of **average fertility, humus rich and well drained**. Good winter drainage is essential to prevent the tubers from rotting.

Tips

Cyclamens make a lovely addition to a woodland or shaded garden. They can also be planted under shrubs and in shaded rock gardens.

Recommended

C. coum forms a small clump of rounded leaves. It grows 2–8" tall and slowly spreads by seed to form a colony of variable size. Flowers in shades of pink and magenta to white are borne from late fall through late winter, later than *C. hederifolium*.

C. hederifolium

C. hederifolium (ivy-leaved cyclamen) forms a low clump of triangular to heart-shaped leaves viened or marbled with silver, gray or bronze markings. It grows 4–6" tall and spreads 6–8". Flowers in shades of pink or sometimes white are produced from fall to early winter and emerge before the new leaves. Plants usually go dormant in summer.

Problems & Pests

Spider and cyclamen mites as well as vine weevils can be a problem. Mice and squirrels may eat tubers or dig them up.

Daffodil
Narcissus

Height: 6–26" **Spread:** 4–12" **Flower color:** white, yellow, orange, peach, pink, bicolored **Blooms:** spring **Zones:** 3–9

POSSIBLY MY FAVORITE PLANT OF ALL, DAFFODILS ARE THE MOST joyful harbingers of spring. They are not only beautiful, versatile and durable, but also disease and critter resistant. Once established, in sun or partial shade, as long as the soil is well-drained, they will return reliably, bloom radiantly and multiply responsibly with little care and great reward. I have a special fondness for the pink-cupped ones. Breeders are getting closer to the elusive red everyday, and now there are ones with pink or yellow in the petals. I also love the "reverse bicolors," those with predominantly yellow perianth and a mainly white cup. They are spectacular in the garden and as cut flowers. And those beguiling miniatures with swept-back petals are simply irresistible!

Planting

Seeding: Species daffodils can be started in summer in a cold frame.

Planting out: Fall

Spacing: 4–8"

Growing

Daffodils grow best in **full sun** or **light, dappled shade.** The soil should be **average to fertile, moist and well drained.** Plant bulbs 2–8" deep in fall, depending on the size of the bulb. The bigger the bulb, the deeper it should be planted. The general rule of thumb is to plant the bulb three times deeper than the height of the bulb. Like cyclamen, daffodils have contractile roots and will eventually adjust to their proper depth level on their own.

Tips

Daffodils are often planted where they can be left to naturalize— planted randomly and informally for a "natural" or "wild" effect—in the light shade beneath a tree, in a woodland or meadow or in mixed beds and borders. Left undisturbed, they will "go forth and multiply" making a more beautiful display each spring. When scattering them in a lawn, however, realize that you should not mow the grass until July, since the daffodil leaves are busy manufacturing food for the bulbs until then. Nevertheless, you can apply growth retardant (available, but not easy to find) to the grass or, if desperate, contact Hertz rent-a-sheep.

To keep your daffodil show looking fresh and to prevent seed production

N. 'Bravoure' (above), *N.* 'Tete-a-Tete' (below)

N. 'Stainless' (above), *N.* 'Lemon Glow' (below)

from sapping the bulb's energy, pull off (deadhead) spent flowers before the ovary (just below the petals) swells. Remove the entire stem, but not the leaves until they turn yellow and begin to wither, usually about July. Hide unsightly foliage among other plants—daylilies and annuals are good camouflage plants to place them among—and refrain from braiding or tying them up with rubber bands. Ouch!

Divide daffodil clumps that have ceased to bloom freely. Conventional practice says to mark them and wait till July, when leaves have thoroughly declined, but it's tough to locate bulbs when leaves are gone. You may move them well before then. Green leaves will flop and look goofy, but don't remove them, as they are the source of the bulb's energy for the next year's growth.

Recommended

Many species, hybrids and cultivars are available. Flowers range in size from $1^{1}/_{2}$–6" across and may be solitary or borne in clusters. There are currently 13 flower-form categories.

N. '**Accent**' grows about 18" tall and bears flowers with crisp white perianths and large, sunproof, upfacing, salmon pink coronas—a vigorous performer.

N. '**Actaea**' grows 16–18" tall and has fragrant flowers with snow white perianths that have broadly overlapping petals and small, flat, yellow coronas edged with bright red. It is the latest flowering of all narcissus and one of the best for naturalizing.

Plant it with Spanish bluebells for a classic combination in a shady spot in your May garden.

N. **'Bravoure'** grows 14–18" tall and bears flowers with white, overlapping petals on its 5" wide perianth and unusually long "stovepipe" of a corona. An award winner and very elegant.

N. **'Decoy'** grows 12–18" high and bears flowers with white perianths and rich, dark coral pink coronas, a stunning and unusual color.

N. **'Ice Follies'** grows 16–18" tall, and bears flowers whose white perianth backs a large, funnel-shaped, frilled yellow corona that fades to white as it matures. It is one of the easiest and most successful bulbs for naturalizing.

N. **'Intrigue'** grows 10–12" tall and is a reverse bicolor daffodil, with a deep yellow perianth and a pale, creamy corona. Quite prominent in the garden.

N. **'Jenny,'** a perennial award-winner, grows 10–12" tall and bears flowers with a gracefully recurved (arched backwards) white perianth and a slender, primrose yellow corona that ripens to creamy white.

N. **'Kaydee'** grows 8–12" tall and bears flowers with pure white, swept-back perianths and vivid, salmon pink coronas.

N. **'Lemon Glow'** grows 18–22" high and has won awards for its unusual, pale primrose color that fades to milky white. Its corona has a ruffly, darker yellow rim. Huge and robust,

Daffodils have triangular stems to help them turn their backs on the breeze, to prevent them from breaking off and to keep them upright. This makes them dance about more than other flowers, including tulips.

it always engenders admiration from garden visitors.

N. '**Mondragon**' grows 12–18" tall and bears flowers with golden yellow perianths, ruffled, tangerine-colored coronas and a subtle apple scent.

N. poeticus '**Plenus**' grows 16–18" tall and bears snowy white, richly fragrant, double white flowers. It is one of the few daffodils that shows up on almost every daffodil list from 1601 through to the catalogs of the early 1900s, and can still be found today.

N. '**Stainless**' grows 18–24" tall and bears flowers with ivory white perianths and luminous white, green-eyed coronas.

N. '**Stratosphere**' is a vibrant, late-blooming jonquil hybrid that grows 26" tall, so it can be planted farther back in the border. It blooms in clusters of three with yellow perianths and orangy yellow coronas. The flowers are remarkably long lasting in the garden and when cut and brought indoors.

N. '**Tete-a-Tete**' is a popular, fragrant miniature, about 6" tall, whose bright yellow flowers in clusters of 1–3 blooms are very early and prolific. Great for naturalizing, they are also easy to force (getting them to bloom before they would normally). In fact, the thousands I grow were purchased in bloom at a supermarket, six to a pot, and planted out after they faded. Nine or ten pots yielded scores of flowers within a few years.

Problems & Pests

Possible problems include nematodes, mites, slugs, large narcissus bulb fly, rot and yellow stripe virus. If spots or blotches with yellow borders develop on leaves, or if flower stalks dry up without producing flowers, control by minimizing moisture on the leaves and flower stalks and by providing good ventilation and light. Destroy any that are heavily infected. Flower blast causes buds to turn brown and dry up before opening. The cause is unknown but is thought to be weather related. Bulbs with this symptom year after year can also be dug and discarded.

N. hybrid (above), *N.* 'Ice Follies' (below)

The cup in the center of a daffodil flower is called the corona, and the group of petals that surrounds the corona is called the perianth.

Daylily
Hemerocallis

Height: 1–6' **Spread:** 2–4' or more **Flower color:** every color except true blue and pure white **Blooms:** summer **Zones:** 2–9

FIREWORKS AREN'T JUST FOR THE FOURTH OF JULY. ALL MONTH long daylilies dazzle. One of the easiest perennials for beginners, they now rival roses in popularity. Modern daylilies come in every color but true white, black or blue, in single shades or intricate color patterns, with varied texture and substance. Blooms may be trumpet-shaped, circular, starry, double, triangular or spidery (very trendy), and as small as violets or a full 9" across. Plants range from 1 to 6' high, some reblooming several times a season. My hands-down favorite is the fragrant, pale yellow species daylily, *H. citrina*. Blooming nocturnally, it's perfect for people who enjoy their gardens at night. Four- to six-foot flower stems challenge marauding deer to reach the flowers. This species is sterile, therefore, self-cleaning—no deadheading—and foliage generally remains unblemished throughout the season.

Planting

Seeding: Not recommended; hybrids and cultivars don't come true to type

Planting out: Spring

Spacing: 1–4'

Growing

Daylilies grow in any light from **full sun to full shade.** The deeper the shade, the fewer the flowers produced. The soil should be **fertile, moist and well drained,** but these plants adapt to most conditions and are hard to kill once established.

Feed your daylilies in spring and mid-summer to produce the best display of blooms. Divide every two to three years to keep plants vigorous and to propagate them. They can, however, be left indefinitely without dividing.

Tips

Plant daylilies alone, or group them in borders, on banks and in ditches to control erosion. They can be naturalized in woodland or meadow gardens. Small varieties are nice in planters. Leaves arrive just in time to hide declining daffodil greenery. Yank off shabby foliage after flowering, and a fresh flush will emerge.

Deadhead small varieties to keep them blooming as long as possible. Be careful when deadheading purple-flowered daylilies because they can stain fingers and clothes.

The petals of daylilies are edible. Add them to salads for a splash of color and a pleasantly peppery taste.

H. hybrid (above & below)

H. hybrid (photos this page)

Recommended

You can find an almost infinite number of forms, sizes and colors in a range of species, cultivars and hybrids. See your local garden center or daylily grower to find out what's available and most suitable for your garden. Several commonly available and attractive daylilies are listed here.

H. 'Autumn Minaret' bears peachy yellow flowers with a faint rosy eye on heavily budded stems up to 6' high in mid-summer and early fall. It grows 5' tall or more.

H. citrina (citron daylily, lemon lily) grows up to 4' tall and spreads about 24". It bears very fragrant, yellow flowers that open in the evening.

H. 'Ed Murray' bears velvety red, ruffled flowers with yellow centers. It grows about 30" tall, with an equal spread.

H. 'Happy Returns' bears yellow flowers for most of the summer. It grows about 16" tall.

H. **'Little Grapette'** has small purple flowers and grows 12" tall and wide.

H. **'Pardon Me'** has bright red flowers with greenish throats. It grows 18" tall and wide.

H. **'Stella de Oro'** was one of the first repeat bloomers introduced. The bright, golden apricot flowers are borne on modest-sized, 12" tall plants. This hybrid will multiply rapidly to fill a long border. Many newer repeat-blooming daylilies in other colors are now available. Plant some for a long show.

H. hybrid (above & center)

Problems & Pests

Problems with rust and slugs are most likely, though problems including *Hemerocallis* gall midge, aphids, spider mites and thrips are possible. Dayliles are a favorite food of deer.

H. 'Stella de Oro' (below)

Dead Nettle

Lamium

Also Called: Lamium, Spotted Dead Nettle, Yellow Archangel
Height: 4–24" **Spread:** indefinite **Flower color:** white, pink, yellow, mauve; plant also grown for foliage **Blooms:** spring, early summer **Zones:** 3–8

TALK ABOUT A BUMMER OF A NAME! FORTUNATELY, DEAD NETTLE bears no actual relationship to prickly nettles but is, instead, a group of useful, color-ful groundcovers. Most are quick-spreading, with the exception of *L. galeobdolon* 'Hermann's Pride,' a first-rate clumping form with a restrained metabolism and beautifully white and green streaked, almost metallic, foliage. Striking even out of bloom, dead nettle's spikes of butter yellow flowers are a bonus Dead nettle grows in a variety of low-light conditions and can take some sun. I grow them as much for the variegated, silvery foliage as for the small, long-lasting, white, pink or mauve flowers. Using this plant throughout a garden can add uniformity, tying areas together. Try *L. maculatum* 'White Nancy' with white impatiens to illuminate a shady spot. Dead nettle also makes a good container plant.

Planting

Seeding: Not recommended; cultivars don't come true to type

Planting out: Spring

Spacing: 12–24"

Growing

Dead nettles prefer **partial to light shade**. They tolerate full sun but may become leggy. The soil should be of **average fertility, humus rich, moist and well drained**. The more fertile the soil, the more vigorously the plants will grow. These plants are drought tolerant when grown in the shade but can develop bare patches if the soil is allowed to dry out for extended periods. Divide and replant in fall if bare spots become unsightly.

Dead nettles remain more compact if sheared back after flowering. If they remain green over winter, shear back in early spring.

L. galeobdolon cultivar (above)
L. maculatum (below)

Tips

These plants make useful groundcovers for woodland or shade gardens or for under shrubs in a border, where they will help keep weeds down.

Keep in mind that dead nettles can overwhelm less vigorous plants. If your dead nettles become invasive, they are easy to remove. Pull some of them up, as needed, making sure to remove the fleshy roots.

Recommended

L. galeobdolon (*Lamiastrum galeobdolon*; yellow archangel) can be quite invasive, though the cultivars are less so. It grows 12–24" tall and spreads indefinitely. The flowers are yellow and bloom in spring to early summer. **'Hermann's Pride'** is a clump former instead of a runner. The leaves are silvery with dark green veins.

L. maculatum 'White Nancy' (above)
L. maculatum 'Beacon Silver' (below)

L. maculatum (spotted dead nettle) is the most commonly grown dead nettle. This low-growing, spreading species grows 8" tall and at least 36" wide. The green leaves often have white or silvery markings. White, pink or mauve flowers are borne in spring and early summer. **'Beacon Silver'** has green-edged, silver foliage and pink flowers. **'Beedham's White'** has yellow to gold foliage with white flowers. **'White Nancy'** bears white flowers and silver leaves with green margins.

Problems & Pests

Rare problems with slugs, powdery mildew, downy mildew and leaf spot are possible.

With their strikingly variegated leaves, these plants are useful for lighting up dark spaces in the garden. They look especially lovely at dusk and in the moonlight.

L. galeobdolon cultivar (above)
L. maculatum cultivar (below)

Dwarf Plumbago

Ceratostigma

Also Called: Plumbago, Leadwort

Height: 10–18" **Spread:** 12–18" **Flower color:** deep blue

Blooms: late summer **Zones:** 5–9

I WAS ONCE TOLD, "THIS IS A PLANT ANY IDIOT CAN GROW." I THINK I have proved that to be true. Should you fail with it, just suppose you are way too intelligent. Chances are, you won't have any problems with this groundcover. It's pretty easy going, its major virtue being the intense electric blue of its flowers. It also has nice red-bronze foliage in fall. The patch I planted 18 years ago has recently started to get a little too confident and is edging its way into neighboring territory. But then, after 18 years, I guess we all start to spread a little more than we should.

This beautiful blue-flowered groundcover originated in China.

Planting

Seeding: Not recommended
Planting out: Spring or fall
Spacing: 12"

Growing

Grow dwarf plumbago in **full sun** or **partial shade**—plants will not bloom in deep, full shade. The soil should be **average** or **rich and well drained**. This quick-growing plant is moderately drought tolerant once established and makes an excellent, tough groundcover. Divide dwarf plumbago in spring.

Dwarf plumbago may not die back completely in winter, and any growth that has been killed or damaged by the cold should be removed in spring. Any unsightly or irregular growth may be removed in fall. Cuttings may be started from new growth in early summer.

Tips

This plant is useful on exposed banks where mowing is impossible or undesirable. Dwarf plumbago also makes a wonderful addition to a rock garden. It creeps happily between the rocks of a stone wall.

C. plumbaginoides (all photos)

Be careful not to disturb the planting site in early spring because the foliage emerges late.

Recommended

*C. **plumbaginoides** (Plumbago larpentiae)* is a low, mounding, spreading plant. Gentian blue flowers appear in late summer. The leathery foliage turns an attractive bronzy red in fall.

Problems & Pests

Powdery mildew causes occasional problems that are preventable with good drainage.

Epimedium

Epimedium

Also Called: Barrenwort, Bishop's Hat

Height: 6–24" **Spread:** 12–30" **Flower color:** white, pink, yellow, red, purple. **Blooms:** mid- to late spring **Zones:** 4–8

TO THE CHINESE, EPIMEDIUM IS "YING YANG HUO," LOOSELY translated as "licentious goat plant," explaining why Western supplement companies have adopted the titillating name by which it is sold in the U.S., "horny goat weed." Its medicinal use, dating back to at least 400 AD, is for treating fatigue and boosting libido. That aside, epimedium is an outstanding choice for any shade garden. Lower growing varieties make superior groundcovers and concealment plants for areas that otherwise might remain barren. They survive in one of the most difficult garden situations: dry shade. The tiny, delicate flowers are charming, the emerging leaves tinged with subtle color, and the plant stays lovely year-round. Critters ignore it. And thank heavens for that. A libido boost is the last thing deer and rabbits need.

Planting

Seeding: Start seed in a cold frame in summer or fall

Planting out: Spring or early fall

Spacing: 12–24"

Growing

Epimedium grows best in **light shade to partial shade** and tolerates full shade. The soil should be **average to fertile, humus rich and moist**, though plants are fairly drought tolerant once established. Cut tattered or withered-looking foliage back in spring before new growth begins.

Tips

These lovely, spring-blooming plants are a popular addition to shade and woodland gardens, as accent plants or groundcovers. They can be planted under taller, shade-providing plants in beds and borders as well as moist, pondside plantings. Epimediums can be slow to establish, but are well worth the wait.

A spray of airy bishop's hat flowers, picked close to the ground when the leaves are mature, will last at least two months in water.

E. x youngianum 'Niveum' (above)
E. rubrum (below)

Recommended

E. alpinum (alpine barrenwort) forms a low clump of bright green, deciduous foliage that turns reddish in fall. It grows 6–12" tall and spreads about 12". It bears yellow flowers with reddish brown sepals. **'Rubrum'** bears dark red or magenta flowers.

E. x *cantabrigense* forms a clump of dark green evergreen foliage. It grows 12–24" tall and spreads about 24". The coppery orange flowers are splashed with red.

E. grandiflorum forms a clump of light green, heart-shaped, deciduous leaves. It grows 8–12" tall and spreads about 12"and bears yellow, white, pink or purple flowers. **'Lilafee'** ('Lilac Fairy') has purple-tinged leaves and purple flowers.

E. x *perralchium* **'Frohnleiten'** is a compact, spreading evergreen plant with reddish green leaves. It grows about 16" tall and spreads 18–24", bearing bright yellow flowers.

E. grandiflorum cultivar (below)

E. x *rubrum* forms a low, spreading clump of reddish green leaves. It grows about 12" tall and spreads 12–18". It bears wine- and cream-colored flowers.

E. x *versicolor* forms a clump of evergreen leaves that emerge a coppery red and mature to green. It grows about 12" tall, with an equal spread. The yellow flowers are sometimes tinged with pink.

E. x *warleyense* forms a spreading clump of evergreen foliage. It grows about 20" tall and spreads up to 30". It bears yellow flowers with scarlet sepals.

E. x *youngianum* forms a clump of deciduous foliage. It grows 8–12" tall and spreads about 12" with white or pink flowers. **'Niveum'** bears white flowers, and the emerging foliage can be quite colorful.

Problems & Pests

Rare problems with vine weevil and mosaic virus can occur.

Epimedium is called bishop's hat because its flowers resemble a clergyman's miter.

E. x *versicolor* cultivar (above)
E. x *versicolor* (below)

Euphorbia

Euphorbia

Also Called: Spurge, Milkwort

Height: 12–24" **Spread:** 12–24" **Flower color:** yellow, green, orange; plant also grown for foliage **Blooms:** spring to mid-summer **Zones:** 4–9

EVER IN A BLISSFUL STATE OF EUPHORBIA, THE INTERNATIONAL Euphorbia Society—trust me, there's a society of *everything* out there—reports there are at least 300 genera and 7500 species in the Euphorbiaceae family. Many are thorny beasts you wouldn't want to mess with in the garden, even if they were hardy here. But, like the cacti some of them emulate, they are exceedingly drought tolerant and animal proof.

I'd never do without *E. amygdaloides* var. *robbiae* (don't panic; you can call it wood spurge) and *E. myrsinites* (donkey-tail spurge.) The former is an indispensable dry-shade groundcover, with deep green rosettes of leaves that have a satiny sheen and typical euphorbic chartreuse yellow flowers, excellent for lighting up dim sites. The latter is a sun-loving succulent, happily weaving in and around other plants, withstanding even the driest, most gravelly, inhospitable soils. It has beautiful, blue-gray leaves and blooms along with the first spring bulbs. A great filler plant.

Planting

Seeding: Start seed in cold frame in spring; use fresh seed for best germination

Planting out: Spring or fall

Spacing: 18"

Growing

Euphorbias grow well in **full sun** and **light shade**. The soil should be of **average fertility, moist, humus rich and well drained**. These plants are drought tolerant and can be invasive in too-fertile soil. They do not tolerate wet conditions.

You can propagate euphorbias by stem cuttings; they may also self seed in the garden. Division is rarely required. These plants dislike being disturbed once established.

Tips

Use euphorbias in a mixed or herbaceous border, rock garden or lightly shaded woodland garden.

If you are cutting the stems for propagation, dip the cut ends in hot water before planting to stop the sticky white sap from running.

You may wish to wear gloves when handling these plants; some people find the milky sap irritates their skin. Euphorbias are also toxic if ingested.

E. dulcis 'Chameleon' (above), *E. myrsinites* (below)

Recommended

E. amygdaloides* var. *robbiae is a very drought-tolerant and shade-tolerant euphorbia. It grows 12–18" tall and spreads about 36". The deep green, glossy leaves are evergreen to semi-evergreen in Zone 5 winters. This variety is excellent for a difficult dry shade garden.

E. dulcis is a compact, upright plant about 12" tall, with an equal spread. The spring flowers and bracts are yellow-green. The dark bronze-green leaves turn red and orange in fall. **'Chameleon'** has purple-red foliage that turns darker purple in fall. It bears clusters of chartreuse flowers. This cultivar and the species will self-seed.

***E.* 'Jade Dragon'** forms a dense, evergreen mound of purple-tinged foliage. It grows about 12" tall, with an equal spread. It bears yellow flowers in early spring.

E. amygdaloides var. *robbiae* (above), *E. myrsinites* (below)

E. myrsinites (donkey-tail spurge, myrtle spurge) develops sprawling stems with succulent, blue-green leaves. It grows 4–8" tall and spreads 12–18". It emerges very early in spring, blooms with clusters of chartreuse bracts and flowers long before the first tulips.

Problems & Pests

Aphids, spider mites and nematodes are possible problems, along with root rot in poorly drained, wet soil.

E. amygdaloides var. *robbiae* (above)
E. dulcis 'Chameleon' (below)

False Solomon's Seal

Smilacina

Also Called: False Spikenard, False Lily-of-the-Valley, Goldenseal, Job's Tears, Solomon's Plume

Height: 36" **Spread:** 2–4' **Flower color:** white **Blooms:** mid- to late spring
Zones: 3–9

OVER 100 YEARS AGO, NATURALIST F. SCHUYLER MATHEWS, WRITING in *Familiar Flowers of Field and Garden*, remarked of this native wildflower's common names—false Solomon's seal, false spikenard and false lily-of-the-valley—"Why should not a plant so deserving have its own good name? We might as well call a Frenchman a false Englishman." He would have been happier, then, I suppose, with goldenseal, Job's tears and Solomon's plume. Undemanding, merely requiring partial shade, false Solomon's Seal is tough, drought tolerant when established and pest resistant. Silken leaves climb its arching stems toward fragrant flowers at the tips. The red fall berries are semi-translucent and last several months unless birds get to them. Quite a worthy garden plant. Maybe Schuyler had a point.

Planting

Seeding: Not recommended; seed has a short period of viability

Planting out: Spring or fall

Spacing: 24"

Growing

False Solomon's seal grows well in **light** or **full shade**. The soil should be of **average fertility, humus rich, acidic, moist and well drained**. Add peat moss to the soil when planting, and mulch with shredded oak leaves or pine needles to provide the acidic, humus-rich conditions this plant enjoys. Divide in spring.

Tips

Use false Solomon's seal in an open woodland or natural garden. In a shaded border it can be combined with hostas and other shade-loving perennials. To get fall fruit, it is best to have several differently named clones. One plant or several plants with the same name don't seem to induce pollinization.

Recommended

S. racemosa forms a spreading clump of upright, arching stems. White, plume-like flowers appear at the tips in spring, followed by berries that ripen in late summer and fall.

Problems & Pests

Rust and leaf spot are possible but rarely serious problems.

Young shoots of both false and true Solomon's seal have long been cooked like asparagus, all of which are in the lily family.

S. racemosa (all photos)

Flowering Fern
Osmunda

Height: 30"–6½' **Spread:** 2"–13' **Flower color:** no flowers, but fertile fronds are a golden to rusty brown **Blooms:** fertile fronds produced in late spring or early summer **Hardiness:** Zones 2–8

NAMED FOR OSMUNDER, THE SAXON GOD OF WAR, *OSMUNDA* IS A genus of about 12 species of vigorous, handsome, distinctive ferns. They differ from other ferns by having large, prominent reproductive spore cases (sporangia) on specialized fronds or parts of fronds, and differ from each other by having the sporangia in different places. For instance, on cinnamon ferns, the cinnamon-colored "wooly" fronds are the fertile ones. (This wool is used by ruby-throated hummingbirds to line their nests.) On the interrupted fern, the sporangia appear in mid-fern, giving the plant its "interrupted" appearance. In the case of the royal fern, the fertile sporangia are at the tip of the fronds. Flowering fern can make a lush and large garden statement if given moist, shady placement.

The "flowers" of flowering ferns are not actually flowers but spore-producing sporangia.

Planting

Seeding: Not recommended
Planting out: Spring or fall
Spacing: 24–36"

Growing

Flowering ferns prefer **light shade** but tolerate full sun if the soil is consistently moist. The soil should be **fertile, humus rich, acidic and moist to wet.** Divide these plants as needed in fall to control their spread or to propagate more plants.

Tips

These ferns will form a large, attractive mass, and should be planted where they will have the room to spread. They make a lovely addition to a moist woodland garden. If you have a water feature, these plants are a good choice for an overflow or bog area, though they can overwhelm smaller plants.

Recommended

O. cinnamonea (cinnamon fern), native to the eastern U.S., has light green fronds that spread out in a circular fashion from a central point. In spring, leafless, cinnamon brown, fertile fronds grow straight up from the center of the plant. It grows 30" to 5' tall and spreads 24–36".

O. claytonia (interrupted fern), also native to the eastern U.S., is similar to cinnamon fern, but has brown, spore-bearing leaflets (pinnae) in the middle of each frond, with sterile green (vegetative pinnae) above and below, giving the frond the effect of being "interrupted" in mid-frond! It grows 2–4' tall.

O. cinnamomea (above)
O. regalis (below)

O. regalis (regal fern) forms a dense clump of fronds 3–6½' tall that spreads 3–13'. The feathery, flower-like, rusty-to-golden brown, fertile fronds stand out among the light green, sterile fronds.

Problems & Pests

Rare problems with rust can occur.

Foamflower

Tiarella

Height: 4–12" **Spread:** 12–24" **Flower color:** white, pink **Blooms:** spring, sometimes to early summer **Zones:** 3–8

FOAMFLOWER, LIKE EPIMEDIUM, IS ONE OF THE BEST FOUR-SEASON shade plants available. These primarily North American natives send up frothy pink or white spring flowers. Their foliage stays attractive all year, certainly in summer. Many have striking dark patterns on their attractively shaped leaves. In winter, foliage develops purple, bronze or red tones as plants huddle together in furry, little mounds. One of the foremost hybridizers is right here in Pennsylvania—a neighbor of mine, in fact. He, and others, keep churning out lovely new cultivars. Plant them with blue woodland phlox, celandine poppy, dark-leaved bugbanes and any shade-loving bulbs.

The starry flowers clustered along the stems look like festive sparklers.

Planting

Seeding: Start seed in cold frame in spring

Planting out: Spring

Spacing: 6–24"

Growing

Foamflowers prefer **partial, light** or **full shade without afternoon sun**. The soil should be **humus rich, moist and slightly acidic**. These plants adapt to most soils. Divide in spring.

Deadhead to encourage re-blooming. If the foliage fades or rusts in summer, cut it partway to the ground. New growth will emerge.

Tips

Foamflowers are excellent groundcovers for shaded and woodland gardens. They can be included in shaded borders and left to naturalize in wild gardens.

Some spread by underground stems and others grow in clumps.

T. cordifolia (above), *T. cordifolia* hybrid (below)

T. wherryii cultivar

Recommended

T. '**Butterfly Wings**' a clump-forming hybrid that combines unusually dissected foliage with great vigor. Another first: the side lobes of most leaves, which are heavily marked with maroon, are raised above the bottom lobe, looking like a butterfly about to take flight. Light pink flowers emerge on 12–14" stems.

T. cordifolia is a low-growing, spreading plant that bears spikes of foamy-looking, white flowers. This species is native to Ohio and eastern North America. It is attractive enough to be grown for its foliage alone, and cultivars with interesting variegation are becoming available. '**Brandywine**' has foliage with purple-red centers. '**Running Tapestry**' has green leaves with bronzy purple veining.

T. '**Green Sword**' is a clump forming hybrid with deeply lobed, pale green leaves. The central lobe is greatly extended, giving the leaf a sword shape. Light pink flowers are produced in spring.

T. '**Jeepers Creepers**' is a spreading hybrid that has maple-shaped leaves with dark brown central markings. Very striking in the garden.

T. '**Maple Leaf**' is a clump-forming hybrid with bronze green, maple-like leaves and pink-flushed flowers.

T. '**Pink Brushes**' is a robust grower with dense spikes of rich pink flowers that last longer than most. The large leaves have an attractive "quilted" texture and are marked with a variable, central, dark blotch.

T. '**Pink Skyrocket**' is an 8" tall, pink-flowered clump with strongly

dissected, shiny green foliage, high-
lighted with a central black blotch.

T. **'Skid's Variegated'** has pastel green
leaves speckled with cream-colored
spots. In winter and early spring, foliage is a magnificent terra cotta color.

T. wherryii (Wherry's foamflower)
forms a slow-growing clump of
heart-shaped, lobed, purple-tinged,
light green foliage. It grows 6–12"
with an equal spread. Star-shaped,
pink-tinged or white flowers are
produced on spikes in spring and
early summer.

Problems & Pests

Rust and slugs are possible
problems.

*Foamflower tea was a Native
American remedy for mouth and
eye ailments.*

T. cordifolia cultivar (above), *T. cordifolia* (below)

Gas Plant

Dictamnus

Also Called: Burning Bush, Dittany, Fraxinella

Height: 16–36" **Spread:** 24" **Flower color:** white, pink, purple
Blooms: early to mid-summer **Zones:** 3–8

I FELL IN LOVE WITH THIS OLD-FASHIONED GARDEN GEM WHEN I first saw a majestic five-foot stand of it growing beside a store I frequented. The owner allowed me to divide the lush planting, so that she and I could transplant some to our home gardens. New to this game, I had no idea what the fragrant, pristine white, regal beauty was. I had no clue that queenly gas plant, having deep, persnickety roots, does not appreciate being disturbed—particularly, I'm sure, being hacked to pieces by a greenhorn gardener. Naturally, the divisions we took died, as did the unfortunate donor plant I'd massacred. Guilt over that episode haunts me still. I have nightmares that when finally I reach the gates of gardener's heaven I'll be turned away, but only after an enraged mob of vengeful Gas Plant Society ghosts flogs me mercilessly with *Dictamnus* wands.

Planting

Seeding: Start seed in a cold frame in late summer or fall

Planting out: Spring

Spacing: 18–24"

Growing

Gas plant grows well in **full sun** and **partial shade**. The soil should be of **average fertility and well drained**. Plants are drought resistant, actually preferring their soil to be on the dry side. They can take a couple of years to establish and may look a bit pathetic until they do. Plants should not be divided because they resent being disturbed. They self-seed, and seedlings can be transplanted while they are still small.

Tips

These tough, drought resistant plants are long-lived and make useful additions to neglected areas of the garden. They also make beautiful and impressive border specimens.

The leaves exude a chemical that reacts with the sun. Touching them may cause photodermatitis, a rash or other skin irritation, on sunny days.

Recommended

D. albus forms a large clump of lemon-scented foliage. It grows 16–36" tall and spreads about 24". Spikes of white- or pink-tinged flowers are produced from early to mid-summer. **'Purpureus'** features light pinkish purple flowers with darker purple veins.

D. albus 'Purpureus' (above & below)

Problems & Pests

Rarely suffers from any problems. Perhaps the various chemicals the plant gives off deter pests and diseases.

Globe Thistle

Echinops

Also Called: Small Globe Thistle

Height: 2–4' **Spread:** 24" **Flower color:** blue, purple **Blooms:** late summer **Zones:** 3–8

ALL AFFILIATES OF THIS SOUTHERN EUROPEAN GENUS HAVE DEEPLY notched, hoary-green foliage and round, spiky, steel blue flowerheads on erect, rigid, branching stems—distinctive qualities for plotting an interesting high-summer garden. The one detriment to dealing with globe thistle is that its leaves have wicked barbs, challenging anyone to touch them. Usually I abstain from planting anything that would dare bite the hand that feeds it, but since globe thistle doesn't require feeding, and rewards me with two full months of carefree good looks in the middle of summer, I make an exception. A superb plant with a strong garden presence, it belongs in mid-border among pinks, yellows and whites. Coreopsis, daylilies and lots of annuals should be in full bloom to keep them company. A great cut flower, it is long lasting and holds its color when dried. I've seen spectacular wreaths made from the flowerheads.

Planting

Seeding: Direct sow in spring

Planting out: Spring or fall

Spacing: 18–24"

Growing

Globe thistle prefers **full sun** but tolerates partial shade. The soil should be of **poor to average fertility and well drained**. Divide in spring when the clump appears dense or overgrown, becomes less vigorous or begins to show dead areas. Wear gloves and long sleeves to protect yourself from the prickles when dividing.

Deadheading prevents self-seeding. Cutting back to the basal foliage after flowering may result in a second round of blooms.

Tips

Globe thistle is a striking plant for the back or center of the border and for neglected areas of the garden that often miss watering.

Recommended

E. bannaticus forms a clump of woolly, grey stems. It grows 2–4' tall and spreads about 24" and bears round clusters of blue or blue-grey flowers in mid- and late summer. 'Taplow Blue' bears bright blue flowers.

E. ritro forms a compact clump of spiny foliage with round clusters of purple or blue 1–2" flowers. It grows 2–4' tall and spreads about 18". **Veitch's Blue** has smaller but more abundant flowers. *E. ritro* **subsp. ruthenicus** is 24–36" tall, has larger flowers and deeply divided basal leaves.

E. ritro (all photos)

Problems & Pests

Globe thistle rarely has any problems, but aphids can show up from time to time.

Goat's Beard

Aruncus

Height: 6"–6' **Spread:** 1–6' **Flower color:** cream, white
Blooms: early summer, mid-summer **Zones:** 3–7

"FROTHY, CREAM-COLORED CLOUDS ATOP SOARING FANS FLOATING above a crown of ferny foliage" hardly sounds like a description for a plant gracelessly dubbed "goat's beard," but that's how a poet might describe this plant. (Well, okay, maybe a *bad* poet.) Resembling, somewhat, an astilbe on steroids, the flowers are said to be sweetly scented. I hadn't noticed this, so I guess I never thought to dab some eau de goat's beard behind my ears. Next time they're in bloom, I'll sniff. Whether in full sun or partial shade, goat's beard likes moisture, but will survive our often droughty summers once its roots are well established. The diminutive kinds are good plants in a woodland border or for edging. The Cherokee supposedly used the roots to make a poultice for bee stings. If they used the giant form, they must have had access to a backhoe.

Planting

Seeding: Use fresh seed and keep soil moist and conditions humid; soil temperature should be 70°–75° F

Planting out: Spring or fall

Spacing: 18"–6'

Growing

This plant prefers **partial to full shade.** If planted in deep shade, it bears fewer blooms. It will tolerate some full sun as long as the soil is kept evenly moist and it is protected from the afternoon sun. The soil should be **rich and moist,** with plenty of **humus** mixed in.

Divide in spring or fall, though goat's beard may be difficult to divide because it develops a thick root mass. Use a sharp knife to cut the root mass into pieces. An axe may be required for large plants with very dense root masses. Fortunately, this plant will remain happy in the same location for a long time.

Goat's beard self-seeds if flowers are left in place, but deadheading maintains an attractive appearance and encourages a longer blooming period. If you want to start some new plants from seed, allow the seedheads to ripen before removing them. You will need to have both male and female plants in order to produce seeds that will sprout. Don't save male flower clusters— they will not produce seeds.

Male goat's beard plants have full, fuzzy flowers; female plants have more pendulous (drooping) flowers.

A. dioicus (above), *A. aethusifolius* (below)

Tips

This plant looks very natural growing at the sunny entrance or edge of a woodland garden, in a native plant garden or in a large island planting. They may also be used in a border or alongside a stream or pond.

Recommended

A. aethusifolius (dwarf Korean goat's beard) forms a low-growing, compact mound. It grows 6–16" tall and spreads up to 12". Branched spikes of loosely held, and cream flowers are produced in early summer. This plant looks similar to astilbe and is sometimes sold by that name.

A. dioicus (giant goat's beard, common goat's beard) forms a large, bushy, shrub-like perennial 3–6' tall, with an equal spread. Large plumes of cream white flowers are borne from early to mid-summer. There are several cultivars, though some can be hard to find. **Var. *astilbioides*** is a dwarf variety that grows only 24" tall. **'Kneiffii'** is a dainty cultivar with finely divided leaves and arching

A. dioicus (below)

stems with nodding plumes. It grows about 36" tall and spreads 18". **'Zweiweltkind'** ('Child of Two Worlds') is a compact plant with drooping, white flowers.

Problems & Pests

No serious pests or diseases.

With their lovely sprays of blossoms, these plants have also been called by the name bride's feathers.

A. dioicus (above), *A. aethusifolius* (below)

Green and Gold

Chrysogonum

Also Called: Goldenstar, Chrysogonum
Height: 6–12" **Spread:** 18–24" **Flower color:** yellow **Blooms:** spring through mid-summer **Zones:** 5–8

REPRESENTED BY ONLY ONE LONELY SPECIES AND A FEW VARIETIES and cultivars, green and gold is nevertheless a very useful, tough, little low-growing groundcreeper. Although it will thrive in shade, I grow mine in full sun in a dry, inhospitable site, even though conventional wisdom says it will only grow in sun if given lots of moisture. This only proves that it's worth experimenting sometimes, trying plants where you need them to grow and seeing what happens. The plants spread politely, not invasively, and the pale, scalloped, green leaves are semi-evergreen in my Zone 6 garden. The prolific, cheery, golden flowers are really delightful, I imagine even more so in shade.

C. virginianum (all photos)

Planting

Seeding: Start seeds in a cold frame in late summer or fall

Planting out: Spring or fall

Spacing: 18–24"

Growing

Green and gold grows well in a sheltered location in **full sun** or **partial shade**. The soil should be of **average fertility, humus rich, moist and well drained**. Plants can be divided in spring or fall. To propagate, rooted stems can be detached from the main plant and transplanted.

Tips

Green and gold is a pretty groundcover, useful in partly shaded woodland gardens and in beds and borders beneath shrubs where it can help suppress weeds.

Recommended

C. virginianum forms a low-spreading groundcover with bright green, slightly hairy leaves. It grows 6–12" tall and spreads 18–24". Plants are dotted with bright yellow, star-shaped flowers from spring through mid-summer.

Problems & Pests

Rarely suffers from any problems.

This woodland plant is native to eastern North America, including Pennsylvania.

Hardy Geranium

Geranium

Also Called: Geranium, Cranesbill

Height: 4–36" **Spread:** 12–36" **Flower color:** white, red, pink, purple, blue **Blooms:** spring, summer, fall **Zones:** 3–8

HARDY GERANIUMS ARE EXPERIENCING A RENAISSANCE OF popularity and with good reason. They are relatively trouble free, adaptable to many situations and suitable for formal and informal landscape schemes. Some prefer full sun, others shade, while some just make themselves at home wherever you put them. Eager to please, many geraniums gladly offer a second flowering if cut back after their first display. Put geraniums in a rock garden, hang them over a wall, edge a border with them. Two of my special pets, *G. sanguineum* and *G. cinereum* 'Ballerina' make excellent groundcovers. Leaf color varies from species to species, and many put on a different show in autumn. Early-blooming *G. macrorrhizum*, with its deliciously citrus-scented leaves, turns bright red for fall. I'm particularly fond of the ones with dark purple foliage.

Planting

Seeding: Species are easy to start from seed in early fall or spring; cultivars and hybrids may not come true to type

Planting out: Spring or fall

Spacing: 12–24"

Growing

Hardy geraniums grow well in **full sun** and **partial** and **light shade**. Some tolerate heavier shade. These plants dislike hot weather. Soil of **average fertility and good drainage** is preferred, but most conditions are tolerated, except waterlogged soil. *G. renardii* needs poor, well-drained soil to grow well.

Divide in spring. Shear back spent blooms for a second set of flowers. If the foliage looks tatty in late summer, prune it back to rejuvenate.

Tips

These long-flowering plants are great in the border, filling in the spaces between shrubs and other larger plants and keeping the weeds down. They can be included in rock gardens and woodland gardens and mass planted as groundcovers.

Recommended

G. **'Brookside'** is a clump-forming, drought-tolerant geranium with finely cut leaves. It grows 12–18" tall and spreads about 24". The deep blue to violet blue flowers appear in summer. (Zones 3–8)

G. **x *cantabrigense*** (Cambridge cranesbill) is a low, trailing plant with fragrant, glossy, evergreen foliage. It

G. sanguineum cultivar (above), *G.* x *oxonianum* (below)

grows 6–12" tall and spreads 12–24". White, pink or purple flowers are borne from spring through summer. (Zones 4–8)

G. *cinereum* (grayleaf geranium) forms a basal rosette of gray-green foliage 4–6" tall and about 12" wide. It produces small clusters of white or pink-veined flowers in early summer. It is often grown in rock gardens and other well-drained spots. **'Ballerina'** is a petite but enduring gem with silver

leaves and pink flowers darkly veined in purple. (Zones 5–8)

G. 'Johnson's Blue' forms a spreading mat of foliage 12–18" tall and about 30" wide. It bears bright blue flowers over a long period in summer. (Zones 3–8)

G. 'Jolly Bee' forms a large mound of foliage. It grows about 30" tall and spreads 30–40", bearing bright blue flowers with white centers from late spring to mid-fall. (Zones 4–8)

G. *macrorrhizum* (bigroot geranium, scented cranesbill) forms a spreading mound of fragrant foliage. It grows 12–20" tall and spreads 16–24". This plant is quite drought tolerant. Flowers in various shades of pink are borne in spring and early summer. **'Album'** bears white flowers in summer on compact plants. **'Bevan's Variety'** bears magenta flowers. (Zones 3–8)

G. x *oxonianum* is a vigorous, mound-forming plant with attractive evergreen foliage; it bears pink flowers from spring to fall. It grows up to 30" tall and spreads about 24". **'A.T. Johnson'** bears many silvery pink

G. pratense (above), *G. sanguineum* var. *striatum* (below)

flowers. **'Katherine Adele'** has variegated green and burgundy foliage and pale pink flowers. **'Wargrave Pink'** is a very vigorous cultivar that grows 24" tall, spreads about 36" and bears salmon pink flowers. (Zones 3–8)

G. pratense (meadow cranesbill) forms an upright clump 24–36" tall and about 24" wide. Many white, blue or light purple flowers are borne for a short period in early summer. This species self-seeds freely. **'Midnight Reiter'** has deep purple leaves and purple-blue flowers. **'Mrs. Kendall Clarke'** bears rose pink flowers with blue-gray veining. **'Plenum Violaceum'** produces purple, double flowers for a longer period than the species because it sets no seed. (Zones 3–8)

G. renardii (Renard's geranium) forms a clump of velvety, deeply veined, crinkled foliage about 12" tall, with an equal spread. A few purple-veined, white flowers appear over the summer, but the foliage is the main attraction. (Zones 3–8)

G. **'Rozanne'** forms a mound of foliage that grows 16–24" tall and spreads 18–24". Violet blue flowers with white centers are borne from late spring to early fall. (Zones 4–8)

G. sanguineum (bloodred cranesbill, bloody cranesbill) forms a dense, mounding clump 6–12" tall and 12–24" in spread. Bright magenta flowers are borne mostly in early summer and sporadically until fall. **'Album'** has white flowers and a more open habit than other cultivars. **'Alpenglow'** has bright, rosy red flowers and dense foliage. **'Elsbeth'** has light pink flowers with dark pink

G. pratense 'Plenum Violaceum' (above)

G. sanguineum (center), G. 'Johnson's Blue' (below)

veins. The foliage turns bright red in fall. **'Shepherd's Warning'** is a dwarf plant to 6" tall with rosy pink flowers. **Var.** *striatum* is heat and drought tolerant. It has pale pink blooms with blood red veins. (Zones 3–8)

Problems & Pests

Rare problems with leaf spot and rust can occur.

Hardy Orchids
Bletilla / Spiranthes

Height: 12–24" **Spread:** 3–24" **Flower color:** pink, white **Blooms:** spring to early summer, fall **Zones:** 5–8

ORCHIDS ARE OFTEN CONSIDERED FRAGILE, TROPICAL RARITIES to be coddled in greenhouses. In reality, except for the driest deserts and wettest aquatic habitats, orchids are found worldwide, including in the arctic tundra. Some 216 species of terrestrial (ground growing) orchids are native to North America. Many are hardy here, including *Cypripedium*, the lady slippers. Most are finicky little devils, however.

In contrast, *Bletilla* and *Spiranthes* are relatively easy for even a gardening neophyte. *Bletilla striata*, from China, Taiwan and Japan, is tough, long blooming and quite beautiful. My eye-level "patch" in a raised bed under a huge tree has hung in there for 15 years and, although I still get only a few 10" stalks each year, I happily accept and appreciate the delicate beauties on their own terms. *Spiranthes* can colonize quickly given a moist site. Its fragrance is often compared to that of vanilla or jasmine.

The name Spiranthes *comes from the Greek words* speira, *meaning "spiral," and* anthos, *meaning "flower." "Nodding ladies' tresses" refers to the nodding habit of the individual florets that make up the flower spike.*

Planting

Seeding: Not recommended

Planting out: Early spring, preferably while dormant

Spacing: 3–18"

Growing

Both species grow best in **partial shade**, in a **sheltered location**. The soil should be **fertile, humus rich, moist and well drained**. Divide plants in early spring while they are dormant.

B. striata (all photos)

Tips

Both of these plants make lovely additions to woodland gardens. *Spiranthes*, if grown in a boggy area, will spread to form an attractive colony. These orchids can also be used in shaded beds and borders as well as next to water features.

Recommended

B. striata forms a clump of deeply veined, long leaves and grows 10–24" tall with an equal spread. It bears delicate sprays of magenta pink flowers in spring and early summer. **Var. *japonica* f. *gebina*** bears creamy white flowers.

S. cernua (nodding ladies' tresses) forms narrow, upright clumps of star-shaped leaves. It sends up spikes of white flowers with yellow centers in fall. It grows up to 24" tall and spreads 3–6". **F. *odorata* 'Chadd's Ford'** has larger, more fragrant flowers than the species.

Problems & Pests

Spider mites, aphids, whiteflies and mealybugs can cause trouble.

Heliopsis
Heliopsis

Also Called: False Sunflower, Ox Eye, Orange Sunflower
Height: 3–5' **Spread:** 18–36" **Flower color:** yellow, gold, orange
Blooms: mid-summer to mid-fall **Zones:** 2–9

THESE EASTERN NORTH AMERICAN NATIVES, WHICH RESEMBLE closely the true sunflowers, make excellent garden plants and provide beautiful flowers for cutting. The one I grow is *Heliopsis* 'Loraine Sunshine'. Although this is supposedly not a long-lived perennial, it does drop seed, which seems to come true to form, so Loraine has been a steady, welcome guest for quite a few years at my place. She's quite the little exhibitionist, too. Her distinctive foliage, white with green netting, is outstanding, an unexpected treat to accompany her brilliant yellow blooms. Butterflies like this plant almost as much as I do. Good companion plants include asters, Shasta daisies, chrysanthemums, phlox and speedwells.

Planting

Seeding: Start seed in the garden in spring once soil temperature is about 68° F

Planting out: Spring

Spacing: 24"

Growing

Heliopsis prefers **full sun** but tolerates partial shade. The soil should be **average to fertile, humus rich, moist and well drained**. Most soil conditions are tolerated, including poor, dry soils. Divide every two or three years in spring or fall. Deadhead to prolong blooming.

Tips

Use heliopsis at the back or in the middle of mixed or herbaceous borders. This plant is easy to grow and popular with novice gardeners.

Recommended

H. helianthoides forms an upright clump of stems and foliage and bears yellow or orange, daisy-like flowers. It grows 3–5' tall and spreads 18–36". **Var.** *scabra*

H. helianthoides (above), *H.* 'Loraine Sunshine' (below)

'**Sommersonne**' ('Summer Sun') bears single or semi-double flowers in bright golden yellow. This cultivar grows about 36" tall.

H. '**Loraine Sunshine**' is about 30" tall and has creamy white foliage with dark green veins. Single flowers are golden yellow.

Problems & Pests

Occasional trouble with aphids and powdery mildew can occur.

Hellebore

Helleborus

Also Called: Lenten Rose, Christmas Rose
Height: 12–18" **Spread:** 12–18" **Flower color:** white, green, pink, purple, yellow **Blooms:** late winter, mid-spring **Zones:** 4–9

NOT ROSES AT ALL, CHRISTMAS AND LENTEN ROSES ARE RELATED instead to buttercups. Unsurpassed for year-round interest, hellebores are revered by beginner and experienced gardeners alike. They are easily grown, quick to mature and long-lived. Leaves are leathery, persistent and highly ornamental. The gently nodding flowers are early, conspicuous yet subtle, and on close inspection, remarkably intricate and beautiful.

Dry powder ground from the root causes violent sneezing. It reportedly has a bittersweet, acrid taste and, like the rest of the plant, is highly poisonous. Despite this, historically it was used as a purgative, to rid children of worms and as part of a long-standing medical tradition to treat mental problems with caustic substances. Pliny the Elder, warning of its toxicity on animals, said "It killeth them." Well, I suppose that's one way to cure a problem.

Planting

Seeding: Not recommended; seed is very slow to germinate and time to flowering size is long

Planting out: Spring or late summer

Spacing: 12–18"

Growing

Hellebores prefer **light, dappled shade in a sheltered site**, but they will accept a fair amount of direct sun if the soil is moist. Corsican hellebore is the best choice for sunny locations. The soil should be **fertile, moist, humus rich, neutral to alkaline and well drained**. Divide in spring, after flowering, whenever plants are becoming too crowded or thinning out in the centers. These plants self-seed and the seedlings are variable.

Protect plants with mulch in winter if they are in an exposed location, though most winters the leaves stay green, and the flowers will poke through the snow in early February. Freshen plants by removing spent leaves in late spring.

Tips

Use these plants in a sheltered border or rock garden, or naturalize in a woodland garden.

All parts of *Helleborus* species are poisonous, and the leaf edges of some species are very sharp, so wear long sleeves and gloves when planting or dividing.

Unlike many plants that look best planted en masse, hellebores make exceptional specimen plants to admire on their own.

H. foetidus (above & below)

H. x hybridus (above), *H. x hybridus* 'Double Pink' (below)

Hellebores make superb cut flowers. Place them high so you can admire their pretty faces.

Recommended

H. argutifolius (Corsican hellebore) forms a large clump of leathery, silvery green leaves and bears clusters of pale green flowers. It grows $1^1/_2$–4' tall and spreads 18–36". (Zones 6–9)

H. foetidus (bearsfoot hellebore, stinking hellebore) is an upright, clumpforming plant. It grows 18–32" tall and spreads 12–18" and bears clusters of fragrant, pale green flowers, often with purple margins. (Zones 5–9)

H. x hybridus (Lenten rose, Oriental hybrids) plants grow about 18" in height and spread. They are very attractive and may be deciduous or evergreen. Plants bloom in late winter to spring in a wide range of flower colors, including white, purple, yellow, green and pink. Almost all plants sold as *H. orientalis* are

H. x hybridus variegated cultivar (above)

hybrids. Trends in breeding include deeper colored flowers, picotees (with differently colored petal margins), doubles and spotted flowers. (Zones 5–9)

H. niger (Christmas rose) is a clump-forming evergreen. It grows 12" tall, spreads 18" and its crystal white flowers bloom in late winter to early spring. (Zones 4–9)

H. odorus has leaves like most of the Oriental hybrids and luminescent, sweet- to musky-scented, soft green, 3" flowers. It is purportedly the best and the toughest of the green-flowered hellebores.

Problems & Pests

Problems may be caused by aphids, crown rot, leaf spot and black rot, and by slugs when the leaves are young.

H. x hybridus (above), *H. niger* (below)

Hens and Chicks

Sempervivum

Also Called: Houseleek

Height: 3–6" **Spread:** 12" to indefinite **Flower color:** red, yellow, white, purple; plant grown mainly for foliage, which can be green, gray or purple
Blooms: summer **Zones:** 3–8

ALMOST EVERYONE REMEMBERS SEEING AND PROBABLY GROWING *Sempervivum* as a youngster. The "hens and chicks" term alone fascinated us, as we could easily observe the baby "chicks" sprouting from the mother plants, probably warping our understanding of where babies come from for many years after. These succulent little plants can grow on almost any surface. Their ability to store water in their thick leaves allows them to live on sunny rocky precipices and in subalpine and alpine places. In the past they were grown on tile roofs, and it was believed they would protect houses from lightning. Grow them in troughs, strawberry pots, living wreaths, in wall crevices, rock gardens, on chair seats (for fun, not for sitting), even on tabletops for a fascinating textural display. There are many cultivars, and a mixed planting can present a colorful, exciting show.

Planting

Seeding: Not recommended; remove and replant young rosettes to propagate

Planting out: Spring

Spacing: 10–12"

Growing

Grow hens and chicks in **full sun** or **partial shade**. The soil should be of **poor to average fertility and very well drained**. Add fine gravel or grit to the soil to provide adequate drainage. Once a plant blooms, it dies. When you deadhead the faded flower, pull up the soft parent plant as well to provide space for the new daughter rosettes that sprout up, seemingly by magic. Divide by removing these new rosettes and rooting them.

Tips

These plants make excellent additions to rock gardens and rock walls, where they will even grow right on the rocks. Never fertilize them, as this encourages bloom and consequently, death.

Recommended

S. arachnoideum (cobweb houseleek) is identical to *S. tectorum* except that the tips of the leaves are entwined with hairy fibers, giving the appearance of cobwebs. This plant may need protection during wet weather.

S. tectorum is one of the most commonly grown hens and chicks. It forms a low-growing mat of fleshy-leaved rosettes, each about 6–10" across. Small, new rosettes are

S. tectorum 'Limelight' & *S. t.* 'Atropurpureum' (above)
S. tectorum (below)

quickly produced that grow and multiply to fill almost any space. Flowers may be produced in summer but are not as common in colder climates.

Problems & Pests

These plants are generally pest free, although some problems with rust and root rot can occur in wet conditions.

Heucherella

x *Heucherella*

Also Called: Foamy Bells

Height: 8–18" **Spread:** 12–18" **Flower color:** white, pink **Blooms:** spring to mid-summer **Zones:** 3–8

ALTHOUGH TOO HIGH-MINDED TO DO SO BY THEIR OWN DEVICES, *Tiarella* and *Heuchera* were first crossbred in 1912. The result and subsequent intergeneric crosses have produced the heucherellas (sounds kind of like a musical group, doesn't it?) with blended characteristics of both parents. As with human breeding, the aim is to garner the best features from both donors. Alas, as with humans, heucherellas seem to have somewhat of a summer heat meltdown problem, even in shade. Other than that, though, they have some nice traits. They're not stoloniferous—that is, the stems do not produce new roots or shoots at the tips—so they make very attractive plants and show great promise with continued hybridizing. All heucherellas are sterile and, therefore, prolific repeat bloomers. Two popular ones are 'Sunspot' and 'Stoplight,' both with deep red hearts on Day-Glo™ yellow leaves, great for illuminating a shadowy path.

Planting

Seeding: No seeds; plants are sterile

Planting out: Spring or fall

Spacing: 12–18"

Growing

Heucherellas grow well in **full sun, partial and light shade, but tolerate full shade.** Their preference seems to be morning sun and afternoon shade. The soil should be **fertile, humus rich, neutral to acidic, moist and well drained.** They don't perform as well when they have to compete with other heavy feeders and when their soil is allowed to dry out for prolonged periods during hot weather. Plants can be divided in spring or fall.

Tips

Heucherellas make a lovely addition to moist borders and open woodland gardens. The brightly colored leaves and low growth habit makes them a much sought-after groundcover plant and the foliage contrasts attractively with a wide variety of other perennials.

Recommended

x _H._ 'Burnished Bronze' forms a mound of bronzy purple leaves. Dark purple stems contrast with the light pink flowers. It grows 8–18" tall and spreads about 18".

x _H._ 'Kimono' has deeply lobed, silvery green leaves with mottled purple veining and bears pale pink flowers. It grows 8–18" tall with an equal spread.

x _H._ 'Stoplight' has chartreuse leaves with red centers. Flowers are

x _H._ 'Sunspot' (above & below)

white. Plants grow 12–18" tall, with an equal spread.

x _H._ 'Sunspot' has yellow-green foliage with red, central veining. Flowers are pink. Plants grow 8–18" tall and spread about 18".

Problems & Pests

Rarely suffers from any problems.

Hibiscus

Hibiscus

Also Called: Rose Mallow

Height: 3–10' **Spread:** 3–4' **Flower color:** red, white, purple, pink, yellow
Blooms: summer, fall **Zones:** 5–9

NOT QUITE AS COLORFUL AS ITS TROPICAL COUNTERPART, BUT
flashy enough in its own right, hardy hibiscus shines in late summer precisely
when gardens and gardeners are flagging—just the pick-me-up needed to get
us through those doggy days. Dinner plate–sized flowers in luminous white,
shocking pink, radiant red, plum purple and recently, a muted yellow, have
crinkly, quilted or satin-smooth petals. Flowers last only a day, but bloom
non-stop from late summer until frost. The large foliage is deep green and
maple- or heart-shaped. Who could believe okra and cotton had such a
glamorous cousin? Hibiscus, slow to emerge in spring—sometimes as late as
June—likes moist soil, making it perfect for wet spots and low-lying areas.
However, it can perform elsewhere with vigilant watering, including in containers. Swamp hibiscus, for instance, can be grown in 5- to 20-gallon containers of wet soil or even under 6" of water in a water garden. Butterflies
love hibiscus.

Planting

Seeding: Soak seeds for 24 hours before starting them in spring.

Planting out: Spring

Spacing: 36"

Growing

Grow hibiscus in **full sun** or **partial shade**. The soil should be **fertile, humus rich and moist** or even periodically wet. Mulch with compost each spring to keep plants blooming their best, and divide in spring.

Plants are slow to emerge in spring. You may want to mark their location so you don't accidentally dig them up. Trimming plants back by about half in early summer will result in bushier, more compact plants.

Tips

These are lovely and interesting plants to use as specimens or at the back of an informal border. The huge flowers create a stunning focal point when the plants are mixed into a border or pondside planting.

Many hibiscus are native to boggy areas, making them useful in low, damp spots, bog gardens and any moist areas near your water feature.

H. 'Kopper King' (above), H. moscheutos (below)

H. *moscheutos* cultivar (above)
H. 'Old Yella' (below)

Recommended

H. coccineus (scarlet rose mallow) is an upright, shrubby plant that grows 10–12' tall and spreads about 4'. It bears red flowers from summer to early fall.

***H.* 'Crown Jewels'** features dark violet foliage and creamy white flowers with a bright red eye. This cultivar grows only 24" tall and 3–4' wide.

***H.* 'Fireball'** is a bushy perennial with bronzy green leaves. It grows about 4' tall and spreads 3–4', and bears large, dark red flowers in late summer and fall.

***H.* 'Kopper King'** has coppery red, maple-like leaves. It grows 36–40" tall and spreads about 36". The large pale lavender to white flowers have dark red or purple centers that streak into the petals.

H. moscheutos (swamp mallow) features pristine white, 8" flowers

that bloom on 3–4' tall plants. Birds, hummingbirds and bees love this plant. It can grow directly in water or out, if kept moist.

H. **'Moy Grande'** is large and bushy with bright green foliage. It bears very large, red flowers in late summer and fall. This plant grows about 5' tall, with an equal spread.

H. **'Old Yella'** has pale yellow, slightly ruffled, 10" blooms with a red eye zone that emerge on sturdy 5' plants.

H. **'Plum Crazy'** has purple, maple-like leaves and light plum purple flowers with darker purple veins. It grows 3–4' tall and spreads about 36".

Problems & Pests

Hibiscus may develop problems with rust, fungal leaf spot, bacterial blight, *Verticillium* wilt, viruses and stem or root rot. Possible insect pests include whiteflies, aphids,

H. moscheutos cultivar (above & below)

scale insects, mites and hibiscus sawfly (*Atomacera decepta*), a serious pest that can turn leaves quickly into lacy skeletons. Watch for tiny, green worms under the leaves.

Hollyfern

Polystichum

Height: 12–24" **Spread:** 12–36" **Flower color:** none **Blooms:** grown for foliage; does not flower **Zones:** 3–8

WHAT WOULD A WOODLAND GARDEN BE WITHOUT FERNS? Hollyfern (Polystichum) is one of the ten largest genera of ferns—there are about 260 species worldwide. The most common one here is the Christmas fern, a native that can even be grown satisfactorily as a houseplant. In fact, it looks rather like the common household Boston fern. It's undemanding, stays semi-green through winter and is highly adaptable to less than ideal conditions, unlike many of its ferny counterparts. The tassel fern is among the best of evergreen ferns, retaining its fronds through the year, rarely requiring trimming. Crosiers open upward and outward, then tips turn downward, holding this pose several days until the frond fully unfurls— a fern ballet. Invite the neighbors.

Planting

Seeding: Spores can be started in a cold frame

Planting out: Spring or fall

Spacing: 12–24"

Growing

Both Christmas fern and tassel fern grow well in **partial shade to full shade**. The soil should be **fertile, humus rich, moist and well drained**. Trim off any fronds that are looking worn out in spring before the new ones emerge. Divide in spring to propagate your plants or carefully remove offsets from the base of the plant.

Tips

These non-invasive ferns make a lovely addition to a moist woodland garden or shaded water feature. Mass plant them in borders or leave them to naturalize in little-used, shaded areas of the garden.

Recommended

P. acrostichoides (Christmas fern) forms a circular clump of arching, evergreen fronds. It grows 12–18" tall and spreads 12–36". Ideal for the shady rockery, this native fern tolerates drier and sunnier places than most ferns.

P. polyblepharum (tassel fern, Japanese tassel fern) forms a circular clump of glossy, dark green, evergreen fronds. Young fronds are covered in golden hairs when they first begin to unfurl and have a tassel-like appearance. Plants grow 12–24" tall and spread about 24". (Zones 5–8)

Problems & Pests

Rarely suffers from any problems.

P. acrostichoides (all photos)

The use of its evergreen fronds to decorate during the holidays gave Christmas fern its common name.

Hosta

Hosta

Also Called: Plantain Lily

Height: 2–36" **Spread:** 5"–6' **Flower color:** white or purple; plants grown mainly for foliage **Blooms:** summer, early fall **Zones:** 3–8

HOSTAS ARE ROYAL MEMBERS OF THE LILY CLAN, AND RULERS OF the shade garden. There is an ever-growing array to choose from—yellows, blues, every shade of green; varying textures; a myriad of leaf shapes and patterns; teeny ones to gargantuan-sized; sun-tolerant types; and even some with red stems. But hostas have a host of predators. Voles burrow underground and do in the roots while up above slugs gorge on emerging foliage, making Swiss cheese of your most expensive beauties. And then come the rabbits, followed by deer—oh, the deer! They come under cover of darkness, like Carl Sandburg's fog "on little cat feet," and do in the entire plant, flossing their teeth with the flower wands for good measure. A tip: hostas are great container plants for the patio. Then you can guard them from the army of critters that share your enthusiasm. Sip your mint julep with a shotgun handy, if necessary.

Planting

Seeding: Not recommended; they rarely come true to type from seed

Planting out: Spring

Spacing: 1–4'

Growing

Hostas prefer **light** or **partial shade** but will grow in full shade. Morning sun is preferable to afternoon sun in partial shade situations. The soil should ideally be **fertile, moist and well drained**, but most soils are tolerated. Hostas are fairly drought tolerant, especially if given a mulch to help retain moisture.

Division is not required but can be done every few years in spring or summer to propagate new plants.

Tips

Hostas make wonderful woodland plants, looking very attractive when combined with ferns and other fine-textured plants. Hostas are also good plants for a mixed border, particularly when used to hide the ugly, leggy lower stems and branches of some shrubs. The dense growth and thick, shade-providing leaves of hostas make them useful for suppressing weeds.

Although hostas are commonly grown as foliage plants, they are becoming more appreciated for the spikes of lily-like flowers, some of which are fragrant and make lovely cut flowers. Some gardeners, however, find that the flower color clashes with the leaves, which are the main decorative feature of the plant. If you don't like the look of the flowers, feel free to remove them before they open—this will not harm the plant.

Recommended

H. **'Baby Bunting'** is a popular cultivar that grows 6–10" tall and spreads 18–26". It has dark green to slightly

H. 'Gold Standard' (above), *H.* 'Francee' (center)

H. cultivar (below)

H. sieboldiana 'Elegans' (above), *H.* cultivar (below)

bluish green, heart-shaped leaves and bears light purple flowers in early to mid-summer.

H. 'Fortunei Aureomarginata' has yellow-margined leaves and is more tolerant of sun than many cultivars. It grows about 23" tall and spreads about 52".

H. 'Fragrant Bouquet' has bright green leaves with creamy yellow margins. The late-blooming (August-September) white flowers are very fragrant. It grows about 18" tall and spreads about 4'.

H. 'Francee' is a good landscaping hosta with puckered, dark green leaves with a narrow, white margin. It grows about 21" tall and spreads about 50".

H. 'Gold Standard' is a hosta fancier's favorite with bright yellow leaves that have a narrow, green margin. It grows about 22" tall and spreads about 5'.

H. 'Halcyon' forms a dense mound of blue-green leaves and bears light purple flowers. It grows up to 18" tall and spreads up to 40".

H. 'June' has bright yellow leaves with blue-green margins and bears light purple flowers. Always one of the top five among hosta devotees, it grows 12–18" and spreads up to 36".

H. 'Krossa Regal' has blue-gray leaves and light purple flowers. It grows up to 32" tall and spreads 3–5'.

H. plantaginea (fragrant hosta) has glossy, bright green leaves with distinctive veins; it grows 18–30" tall, spreads to about 36" and bears large, white, fragrant flowers in late summer.

H. 'Royal Standard' is durable and low growing. It reaches only 4–8" in height and spreads up to 36". The dark green leaves are deeply veined, and the highly fragrant flowers are light purple to off white.

H. **'Sagae'** forms an upright mound of gray-green leaves with with yellow margins. It bears light purple flowers. It grows about 32" tall and spreads up to 6'.

H. sieboldiana (Siebold's hosta) forms a large, impressive clump of blue-green foliage. It grows about 24" tall and spreads up to 65". The early summer flowers are a light grayish purple that fades to white. **'Elegans'** has deeply puckered, blue-gray foliage and light purple flowers. It was first introduced to gardens in 1905 and is still popular today. It grows about 24" tall and spreads up to 65".

H. **'Sum and Substance'** has immense gold-green leaves and may spread to 6' with a height of 36". This cultivar is slug resistant.

H. **'Tokudama'** grows slowly into a groundcovering clump about 18" tall and 44" wide. It has blue, heart-shaped, puckered leaves and produces white flowers in early to mid-summer. The foliage is slug resistant.

Problems & Pests

Slugs, snails, leaf spot, crown rot and chewing insects such as black vine weevils are all possible problems for hostas. Varieties with thick leaves tend to be more slug resistant. Voles and deer—oh, dear!—can discourage even the most avid hosta fanatic.

H. 'Royal Standard' (above), *H.* 'Francee' (below)

Indian Pink

Spigelia

Also Called: Pinkroot

Height: 12–24" **Spread:** 18" **Flower color:** red and yellow bicolored
Blooms: spring and early summer **Zones:** 6–9

A NEAT AND DAINTY PLANT, YET A TRUE TRAFFIC STOPPER IN THE flower border, there is much to recommend indian pink. In my garden, multiple tubular, scarlet blossoms with brilliant yellow starbursts at the tip shoot upright on approximately 18" stems, blooming in succession from the bottom up over a long period. Dark green foliage persists all season. *Spigelia marilandica* is an easy-to-grow, underused native that fares well just about anywhere in the garden. In open woods it can form large colonies, but it is much tamer in an average garden setting. Indian pink was voted one of the top ten hummingbird plants by Operation Rubythroat (rubythroat.org), an international project studying the behavior and distribution of the ruby-throated hummingbird. Indian pink root was used by Native Americans to treat worms, but considering it has close cousins in the strychnine family, I'd say, "don't try this at home."

Planting

Seeding: Easily grown from seed; sow directly in the garden in late summer or fall

Planting out: Spring

Spacing: 12–18"

Growing

Indian pink prefers **light** or **partial shade** but will tolerate full sun if the soil remains moist. The soil should be **fertile, moist and well drained**. Divide plants in spring.

Tips

Indian pink makes a nice addition to a sunny or shaded bed, border or woodland garden. It can be included in wildflower and native plant gardens. Try it with white bleeding hearts, ferns, columbines, mayapples, or with anything that thinks it can outshine this attention grabber.

S. marilandica (above & below)

Recommended

*S. **marilandica*** forms an upright clump. It grows about 24" tall and spreads about 18". The tubular flowers are cherry red on the outside and vivid yellow on the inside.

Problems & Pests

Rare problems with leaf spot or powdery mildew can occur. No serious pests. Deer don't seem to be interested. Perhaps they're afraid of hummingbird vengeance.

Indian pink is one of the best plants for attracting hummingbirds to the garden.

Iris

Iris

Height: 4"–4' **Spread:** 6"–4' **Flower color:** many shades of pink, red, purple, orange, blue, white, brown, yellow **Blooms:** spring, summer, sometimes fall **Zones:** 3–10

OLYMPIAN MESSENGER, FED-EX AGENT TO THE GODS, THE GODDESS Iris used the rainbow as her route between heaven and earth. Rainbow pieces adhered to her feet, so that wherever she walked on earth, her footprints left behind multicolored flowers, some 300 species and countless cultivars, all of which you will want to try in your garden. One of my favorites is spuria iris, with its smooth, stately, sword-like leaves. Perched like butterflies, fragrant white flowers with yellow splotches on the falls alight on 4–5' tall stems. Years ago, someone gave these to me from an old country farm property. They were probably once more popular than they are today, but certainly deserve more promotion. They are decidedly regal, trouble-free, have perfect upright posture, grow under a variety of less-than-ideal situations and make stunning cut flowers.

Planting

Seeding: Not recommended; germination is erratic and hybrids and cultivars may not come true to type

Planting out: Late summer or early fall

Spacing: 2"–4'

Growing

Most irises prefer **full sun** and tolerate light or dappled shade. The soil should be **average to fertile and well drained**. Japanese iris and Siberian iris prefer a moist but still well-drained soil.

Divide in late summer or early fall. Bearded iris and variegated iris must be divided every few years to maintain good condition. When dividing bearded iris rhizomes, replant with the flat side of the foliage fan facing the garden. Dust the toe-shaped rhizome with a powder cleanser before planting to help prevent soft rot.

Deadhead irises to keep them tidy. Cut back the foliage of Siberian iris in spring.

Tips

All irises are popular border plants, but Japanese iris and Siberian iris are also useful alongside a stream or pond, and dwarf cultivars make attractive additions to rock gardens.

Recommended

I. cristata is a woodland iris that prefers full shade, but will tolerate some sun. It forms a low clump of bright green leaves. Plants grow about 4" tall and spread to form a colony 12" or more wide. Mauve

I. cristata (above), *I. tectorum* (below)

The wall of a 3500-year-old Egyptian temple features an iris, making this plant one of the oldest cultivated ornamentals.

I. hybrid 'Before The Storm' (above)

I. hybrid 'Eyebright' (center), *I. sibirica* (below)

flowers have white patches at the center. Variations with different flower colors are available.

I. ensata (*I. kaempferi*; Japanese iris) is a water-loving species. It grows up to 36" tall and spreads about 18". White, blue, purple or pink flowers are borne from early to mid-summer. This species rarely needs dividing and is resistant to iris borers.

I. **hybrids** (*I. germanica*; German bearded iris) have flowers in all colors. *I. germanica* has been used as the parent plant for many desirable cultivars and hybrids, which may vary in height and width from 6" to 4'. Flowering periods range from mid-spring to mid-summer, and some cultivars flower again in fall. These hybrids are drought tolerant.

I. pallida (sweet iris) is rarely grown, but its variegated cultivars are a popular addition to the perennial garden. Light purple flowers are borne on plants that grow about 24" tall and spread 12". **'Argentea Variegata'** has foliage with cream and green stripes. **'Variegata'** ('Aurea Variegata') has foliage with gold and green stripes.

I. reticulata (netted iris) forms a small clump 4–10" tall. This iris bears flowers in various shades of blue and purple. The plants grow from bulbs and can be left undisturbed. Bulbs sometimes divide naturally, and plants may not flower again until the bulbs mature.

I. sibirica (Siberian iris) is more resistant to iris borers than other species are. It grows 2–4' tall and 36" wide and flowers in early summer. Many cultivars are available in various shades, mostly purple, blue and white. Plants take a year or two to recover after dividing.

***I. spuria* ssp. ochroleuca** (Spuria
iris) is a rhizomatous perennial with
handsome, upright, sword-shaped
foliage clumps. It grows 4–5' tall and
spreads 24–36". Plants are topped
with showy, fragrant, white flowers
with yellow blotches in late spring.

I. tectorum (Japanese roof iris) has
broad, glossy, medium green leaves
and bears mauve flowers with lighter
and darker markings. It forms dense
clumps 12–18" tall that spread up to
36". This iris performs best in dappled
shade, but flowers in sun or shade and
is drought tolerant.

Problems & Pests

Irises have several problems that can
be prevented or mitigated through
close observation. Iris borers are a
potentially lethal pest. They burrow
their way down the leaf until they
reach the rhizome, where they continue eating until there is no root
left. The tunnels they make in the
leaves are easy to spot, and if
infected leaves are removed and
destroyed or the borers squished
within the leaf, the borers will never
reach the rhizome.

I. hybrid (above), *I. pallida* 'Argentea Variegata' (below)

Leaf spot is another problem that
can be controlled by removing and
destroying infected leaves. Be sure to
give irises the correct growing conditions. Too much moisture for some
species will allow rot diseases to settle in and kill the plants. Plant rhizomes high in the soil to deter root
rot. Slugs and aphids may also cause
some trouble.

Jack-in-the-Pulpit
Arisaema

Also called: Indian Turnip

Height: 12–24" **Spread:** 6" **Flower color:** purple, green **Blooms:** spring, early summer **Zones:** 4–9

MY LAWNMOWING GUY, WHO HAS SEEN JUST ABOUT EVERYTHING, almost plowed into a tree, head spinning nearly full circle, when he spotted a lone *A. sikokianum* blooming in the garden. So stunning is this member of a fascinating genus of garden wonders (the plant, that is, not the lawn guy), that only those who have passed a recent EKG exam should be allowed to view them. There are many other species akin to our understated native, *A. triphyllum*, that are still more bizarre looking, all of which will assuredly generate lively conversation.

Calcium oxalate crystals permeating all parts of this plant cause severe burning if ingested. So caustic are they, in fact, that they can be lethal. Nevertheless, Native Americans cooked and dried the tubers, nullifying these effects, and used them as vegetables. They also used the tubers medicinally for cold symptoms, sore eyes and skin infections. The tuber has also been used for making laundry starch.

Planting

Seeding: Start seeds in a cold frame
in spring or fall. From seed, depending on species, expect to wait two to
four years to have a flowering-sized
plant. Size of the seed determines
the amount of food reserves and,
consequently, how large the plants
will grow during the first season.

Planting out: Spring or fall

Spacing: 6"

Growing

Jack-in-the-pulpit grows well in
partial or **light shade** and prefers
cool conditions. The soil should be
**fertile, loose, neutral to acidic,
moist and, especially, well drained**.
This plant requires shade in summer, and regular watering until late
summer, but it can be dry from late
summer through to the following
spring. This tuberous perennial
doesn't need to be divided, but the
small offsets that grow at the base of
the plant can be transplanted to
propagate more plants. Plant corms
of most species five or six inches
deep, dwarf species about three
inches deep. Mulch the planting area
at least for the first winter.

*The flowers consist of a spathe, the
showy outer hood and the spadix,
the flower spike, which is held
within the spathe.*

A. triphyllum (above), A. sikokianum (below)

Tips

Jack-in-the-pulpit makes a great addition to the woodland garden and is sure to be a conversation starter. Though each plant produces only a few leaves and a single flower, a group of them will make a definite impact, and as the offsets begin to fill in, small patches or colonies of plants will establish themselves. Include them in shaded borders, particularly those in moist spots on the north side of the house where the conditions will be cooler.

Recommended

A. ringens produces two lobed leaves and a single large flower with green and purple stripes in early summer followed by showy clusters of red berries in fall. It grows about 12" tall. (Zones 6–9)

A. sikokianum (above), *A. ringens* (below)

Some Arisaema *are male, some female, some both, and some change back and forth* (paradioecious*). Usually, those that do are male when young, become female when they gather enough energy to reproduce and, a year after doing so (fruiting), often revert to being male.*

A. sikokianum produces two lobed leaves, sometimes with silvery variegations and a single large flower with deep maroon (almost black) and white stripes and a luminescent, white, light bulb–like spadix (60 watts, at least!) in spring followed by showy clusters of red berries in fall. It grows 12–18" tall. (Zones 5–9)

A. triphyllum produces one or two lobed leaves and a pale green- or purple-striped flower in spring or early summer followed by showy clusters of red berries in fall. It grows 6–24" tall. (Zones 4–9)

Problems & Pests

Slugs can cause trouble and a few leaf diseases, such as a rust fungus, can occur.

A. triphyllum (above), *A. ringens* (below)

Jacob's Ladder

Polemonium

Height: 8–36" **Spread:** 8–16" **Flower color:** purple, white, blue
Blooms: late spring, summer **Zones:** 3–7

FRANKLY, I HAVE NOT HAD MUCH SUCCESS WITH JACOB'S LADDER, but don't let that stop you from trying a plant with which others have great luck. (I'm sure Jacob hasn't been thrilled with me, either.) The plant has been grown in gardens since the 16th century and probably long before. The prettiest ones, those with variegated foliage, have been recorded since the 18th century. The newer cultivars are getting rave reviews. Maybe I'll get brave and try again. Our native species, *Polemonium reptans*, is also known as "abscess root." For centuries it has served as an astringent, expectorant and to treat the bites of venomous snakes and insects.

Planting

Seeding: Start seed in spring or fall. Keep soil temperature at about 70° F. Seed can take up to a month to germinate.

Planting out: Spring

Spacing: About 12"

The leaflets of the foliage are organized in a neat, dense, ladder-like formation, giving these plants their common name.

Growing

Jacob's ladder species grow best in **partial shade** or **very light shade**. The soil should be **fertile, humus rich, moist and well drained**. Deadhead regularly to prolong blooming. These plants self-seed readily. Division is rarely required but should be done in late summer if desired.

Tips

Include Jacob's ladder plants in borders and woodland gardens. *P. caeruleum* can be used as a tall focal point in planters. *P. reptans* can be used in rock gardens and as an edging along paths.

Recommended

P. caeruleum is the commonly grown Jacob's ladder. This plant grows 18–36" tall and spreads about 12". It forms a dense clump of basal foliage, with leafy upright stems that are topped with clusters of purple flowers. **'Apricot Delight'** produces many mauve flowers with apricot pink centers. **'Snow and Sapphires'** has dark green leaves edged with creamy white and deep purple-blue flowers.

P. reptans (creeping Jacob's ladder) is a very hardy, mounding perennial 8–16" tall, with an equal spread. It bears small, blue or lilac flowers in late spring and early summer. **'Stairway To Heaven'** is a vigorous plant with green leaves that are broadly edged in creamy white with short spikes of blue flowers.

P. yezoense **'Purple Rain'** has dark, bronzy purple foliage. It produces clusters of large blue flowers for a

P. caeruleum (above), *P. caeruleum* cultivar (below)

long period in summer. It grows about 24" tall, with an equal spread.

Problems & Pests

Powdery mildew, leaf spot and rust are occasional problems.

Joe-Pye Weed
Eupatorium

Also called: Boneset, Snakeroot
Height: 2–7' **Spread:** 1–4' **Flower color:** white, purple, blue, pink
Blooms: late summer, fall **Zones:** 3–9

WILDFLOWER ENTHUSIAST FORMER U.S. SENATOR GEORGE D. AIKEN called Joe-Pye weed "a good natured lummox, willing to grow anywhere for anyone." It is often a staple, along with native goldenrods, sunflowers and tall white asters, in a late summer to early fall hedge or meadow. Although *Eupatorium* has many medicinal properties, *E. rugosum*, the attractive, white-flowered form commonly called white snakeroot, is highly toxic if eaten. Before this was discovered, cows fed the plants gave tainted milk that may have killed thousands of Americans in the eastern half of the country, including Abraham Lincoln's mother, Nancy Hanks. This species still affects grazing live-stock and is not recom-mended as an ornamental.

Planting
Seeding: Start seed indoors in late winter or early spring; soil temperature should be 59°–68° F
Planting out: Spring
Spacing: 18–36"

Growing

Joe-Pye weed prefers **full sun** but tolerates partial shade. *E. dubium* 'Little Joe' is the cultivar most tolerant of partial shade. The soil should be **fertile and moist**. Wet soils are tolerated. Divide plants in spring when clumps become overgrown. Don't put off dividing if space is a problem, because dividing oversized clumps is a tough job.

Pruning growth back in May encourages branching and lower, denser growth, but it can delay flowering.

Tips

This plant can be used in a moist border or near a pond or other water feature. The tall types are ideal in the back of a border or at the center of a bed, where they will create a backdrop for lower-growing plants.

It may take a couple of seasons for Joe-Pye weed to mature, so don't crowd it.

Recommended

E.dubium 'Little Joe' is a dwarf plant when compared to other Joe-Pye weeds, growing about 4' tall and about 36" wide. It bears clusters of mauve or lavender flowers in mid-summer.

E. purpureum ssp. *maculatum* is a huge plant 5–7' tall and 3–4' wide. In late summer it bears clusters of purple flowers and is difficult to tell apart from the closely related *E. purpureum*. 'Gateway' is slightly shorter, growing up to 6' tall. The large flower clusters are rose pink and the plant's stems are reddish.

E. purpureum ssp. maculatum

The flowers attract butterflies to the garden and can be used in fresh arrangements.

Problems & Pests

These plants may have occasional problems with powdery mildew, fungal leaf spot, rust, aphids, whiteflies and leaf miners.

Lady's Mantle
Alchemilla

Height: 3–18" **Spread:** 20–24" **Flower color:** yellow, green
Blooms: summer, early fall **Zones:** 3–7

IT WAS LOVE AT FIRST SIGHT WHEN I VISITED ENGLISH GARDENS and saw lady's mantle everywhere, ebullient chartreuse blossoms foaming over the edge of walkways under the feet of roses, delphinium and every other delightful British beauty that never looks as lovely back here. I was thrilled that this incredible plant turned out to be so effortless to grow in Pennsylvania, as so many plants that do well there fail in our very different climate. Lady's mantle is perfection in and out of bloom. Its flower color and texture complement (and compliment) all its "bed partners"; its scalloped leaves are always fresh; it tolerates drought; it is unappealing to pests; and is totally undemanding. Moreover, watching droplets of water or dew bead up and swirl on the felted leaves can be considered garden entertainment.

Planting

Seeding: Direct sow fresh seed or start in containers in fall or spring; transplant while seedlings are small

Planting out: Spring

Spacing: 24"

Growing

Lady's mantle plants prefer **light** or **partial shade**, with protection from afternoon sun. They dislike hot locations, and excessive sun may scorch the leaves. The soil should be **fertile, humus rich, moist and well drained**. These plants are drought resistant once established. Division is rarely required but can be done in spring before flowering starts or in fall, once flowering is complete. If more plants are desired, move some of the self-seeded plants that are bound to show up to where you want them. Deadhead to keep things looking tidy and prevent excessive re-seeding. For *A. mollis*, deadheading may encourage a second flush of flowers in late summer or fall.

A. mollis (above & below)

Tips

Lady's mantles are ideal for grouping under trees in woodland gardens and along border edges, where they soften the bright colors of other plants. A wonderful location is alongside a pathway that winds through a lightly wooded area. They are also attractive in containers.

If your lady's mantle begins to look tired and heat stressed during summer, rejuvenate it in one of two ways. Trim the whole plant back, encouraging new foliage to fill in, or

remove the dead leaves and then trim the plant back once new foliage has started to fill in. Leave plants intact over the winter, then clean them up in spring.

Recommended

A. alpina (alpine lady's mantle) is a diminutive, low-growing plant that reaches 3–5" in height and up to 20" in spread. Soft white hairs on the backs of the leaves give the appearance of a silvery margin around each leaf. Clusters of tiny, yellow flowers are borne in summer. This species is an excellent trough plant in combination with miniature hostas. It is not nearly as vigorous or long-lived as *A. mollis,* but it is beautiful and worth growing, nevertheless.

A. mollis (above), *A. alpina* (below)

*A. **mollis*** (common lady's mantle) is the most frequently grown species. It grows 8–18" tall and spreads up to about 24". Plants form a mound of soft, rounded foliage, above which are held sprays of frothy-looking, yellowish green flowers in early summer. This species is extremely drought tolerant.

Problems & Pests

Lady's mantles rarely suffer from any problems, though fungi may be troublesome during warm, wet summers. These plants are deer resistant.

The airy flowers, although rather unpleasant smelling, make a fabulous filler for fresh and dried flower arrangements

A. alpina (above), *A. mollis* (below)

Lamb's Ears

Stachys

Also called: Woolly Betony
Height: 6–18" **Spread:** 18–24" **Flower color:** pink, purple
Blooms: summer **Zones:** 3–8

LAMB'S EARS PERK UP ALL OVER MY GARDEN, AND I NEVER PLANTED a single one. Birds (or elves) brought them in, and I'm not too proud to accept donations. Their silver color is a great foil for any color scheme, and the velvet foliage is downright sensuous! Who can resist rolling a leaf between their fingers? In fact, if you see me rolling around on a carpet of these, please don't report me to the authorities. The straight species blooms and attracts hordes of bees. They are pleasantly scented, but I try to deadhead them before they get out of hand. Avoid this task by planting cultivars that don't flower at all. There are other *Stachys*, such as *S. macrantha*, that are not fuzzy, not gray and not called lamb's ears, but make excellent, non-bleating garden assets you will be pleased to add to your herd.

Planting

Seeding: Direct sow or start in containers in cold frame in spring

Planting out: Spring

Spacing: 18–24"

Cut lamb's ears' pleasantly scented flowerheads when they are in bud or after they bloom, and hang them to dry for use in dried-flower arrangements.

Growing

Lamb's ears grow best in **full sun**. The soil should be of **poor to average fertility and well drained**. The leaves can rot in humid weather if the soil is poorly drained. Divide in spring.

Remove spent flower spikes to keep plants looking neat. Select a flowerless cultivar if you don't want to deadhead. Cut back diseased or damaged foliage; new foliage will sprout when the weather cools down.

Tips

Lamb's ears make a great groundcover in a new garden where the soil has not yet been amended. It can be used to edge borders and pathways, providing a soft, silvery backdrop for more vibrant colors in the border. For a silvery accent, plant a small group of lamb's ears in a border.

Leaves can look tatty by mid-summer. The more of this plant you use, the more you will have to clean up. Plant only as much as you will tend, or plant in an out-of-the-way spot where the stressed foliage will not be as noticeable.

Recommended

S. byzantina (*S. lanata*) forms a mat of thick, woolly rosettes of leaves. Pinkish purple flowers are borne all summer and attract bees. **'Big Ears'** ('Helen von Stein') has greenish silver leaves that are twice as big as those of the species. **'Silver Carpet'** has silvery white, fuzzy foliage; it rarely, if ever, produces flowers.

S. byzantina 'Big Ears'

Plant this cultivar beneath alliums, yarrow, geranium, peonies or iris that like dry conditions.

S. macrantha **'Robusta'** ('Superba'; big betony, hedge nettle) is the best and most commonly sold form of lamb's ears. Crinkled, dark green scalloped leaves form 10" high mats that spread 24", rapidly if given moist soil. This cultivar tolerates dry soil, as well. Spikes of densely packed, rosy pink to lavender flower clusters rise 8" above foliage. Plant in sun to part shade. Plant this with chartreuse-flowered plants such as lady's mantle for a very striking display. *S. macrantha* **'Alba'** has pure white flowers.

Problems & Pests

Fungal leaf problems including rot, powdery mildew, rust and leaf spot are rare but can occur in hot, humid weather. *S macrantha* is less susceptible to these.

Ligularia

Ligularia

Height: 3–6'   **Spread:** 2–5'   **Flower color:** yellow, orange; ornamental foliage   **Blooms:** summer, sometimes early fall   **Zones:** 4–8

IF YOU'RE LOOKING FOR BOLD, DRAMATIC PLANTS WITH STRONG vertical accents for a partially shaded bog garden or a moist woodland area that gets a reasonable amount of sun, ligularias might be ideal. They can be tricky to grow, though, not to mention hard to place anywhere else in the garden because they like cool temperatures and need moist soil and shade, or the foliage wilts during the heat of the day. Yet they need enough sun to encourage the spectacular spires of bloom. In less than ideal situations, you can disguise the foliage's fainting spells by interplanting it with large hostas. But if you have a dry garden, you'll be setting yourself up for disappointment with these plants.

Planting

Seeding: Species can be started outdoors in spring in containers; cultivars rarely come true to type

Planting out: Spring

Spacing: 2–5'

Growing

Ligularias should be grown in **light shade** or **partial shade** with protection from the afternoon sun. The soil should be of **average fertility, humus rich and consistently moist**. Division is rarely, if ever, required but can be done in spring or fall to propagate a desirable cultivar.

Tips

Use ligularias alongside a pond or stream. They can also be used in a well-watered border or naturalized in a moist meadow or woodland garden.

The foliage can wilt in hot sun, even in moist soil. The leaves will revive at night, but this won't help how horrible they look during the day. If your ligularia looks wilted, it is best to move the plant to a cooler, more shaded position in the garden.

Recommended

L. dentata (bigleaf ligularia, golden groundsel) forms a clump of rounded, heart-shaped leaves. It grows 3–5' tall and spreads 3–4'. In summer and early fall it bears clusters of orange-yellow flowers, held above the foliage. These plants can survive temperatures as low as -30° F. **'Britt Marie Crawford'** has bronzy purple foliage that contrasts with the clusters of bright yellow flowers.

L. dentata (above), *L.* 'The Rocket' (below)

***L.* 'The Rocket'** has heart-shaped leaves with ragged-toothed margins. The leaf veins are dark, becoming purple at the leaf base. It grows 4–6' tall.

Problems & Pests

Ligularias have no serious problems, though slugs can damage young foliage.

Lily

Lilium

Height: 3–8' **Spread:** 12–36" **Flower color:** white, red, pink, purple, yellow, orange **Blooms:** early, mid or late summer **Zones:** 4–8

LILY IS QUEEN OF THE SUMMER PERENNIAL BORDER. FRAGRANT, colorful and varying in form, lilies are available in an almost infinite selection. Recent creations are Oriental-trumpet hybrids, called orienpet lilies. By selecting a combination of early, mid-season and late-blooming cultivars, you can have lilies in flower from mid-June through late September with only minimal care. Don't overlook the awe-inspiring natives, *Lilium canadense* and *L. superbum*, with nodding lanterns above whorled-leaf dressed stems up to 10' feet high. Other species lilies are easy, garden-worthy selections, but my all-time favorite is Formosa lily grown from seed in my first year of gardening. For over 20 years these spectacular, perfumed white trumpets have loomed some nine feet tall, well above deer-lip level, over my late August/ September garden. After the flowers fade, the seed pods turn upward, opening elegantly at the tips to form winter-proof candelabra as garden ornaments, scattering seeds on every breeze. Seed pods can also be used in dried arrangements.

Planting

Seeding: Start seed in spring in a cold frame or under grow lights indoors

Planting out: Fall

Spacing: Hybrids 6–12"; species 12–36"

Growing

Lilies grow best in **full sun** but like to have their roots shaded. The soil should be **fertile, humus rich, moist and well drained.** Most bulbs should be planted two to three times deeper than the bulb is tall. Plants may need staking, especially in windy or poorly sheltered locations. Peony hoops installed when shoots are young may be sufficient to hold up mature plants. To propagate more plants, remove scales, or transplant offsets in fall once the parent plant goes dormant. Some lilies form tiny bulbils along the stem that can be removed and planted to reap more lilies.

Tips

Lilies are often grouped in beds and borders and can be naturalized in open woodland gardens and near water features. These plants are narrow but tall; plant at least three together to create some volume.

Recommended

L. 'African Queen' is a large, vigorous trumpet lily that bears clusters of yellow to apricot-orange, trumpet-shaped flowers. Plants grow 4–6' tall.

L. 'Black Beauty' is a vigorous Oriental lily that yields clusters of dark purple-red flowers with recurved petals. Plants grow up to 5' tall.

L. hybrid (above), *L.* Asiatic lily (center)

L. species (below), *L.* 'Stargazer' (opposite)

L. hybrid (this page)

L. canadense (Canada lily) is a magnificent species lily that produces a candelabrum of slender, pendant, orange, red or yellow, bell-shaped flowers in July. Plants grow 3–6' tall and spread by stolons from which new bulbs form. Under the ideal conditions of rich, moist soil and sun to partial shade, these slow-to-start plants may spread to form an attractive colony. The flowers attract hummingbirds. Plants may flop over unless they are planted in a sheltered location or support is provided.

L. 'Citronella' is a vigorous Asiatic lily that produces up to 30 bright yellow flowers that are speckled with reddish dots. Plants grow 4–5' tall.

L. formosum (Formosa lily) is an elegant species lily that grows 2–9' tall. It produces fragrant, white flowers flushed purple on the outsides, singly or in clusters of up to 10 blooms. This cultivar rarely requires staking. Some bloom the first year from seed.

L. martagon (martagon lily, turk-scap lily) forms a clump of vigorous stems, growing 3–6' tall. The small, bright pink flowers with their recurved petals may have dark speckles and are borne in clusters of up to 50 blooms.

L. regale (regal lily) forms an upright to arching clump of stems growing 2–6' tall and develops clusters of white, trumpet-shaped flowers that are flushed light purple on the outsides.

L. 'Scheherazade' is a vigorous orienpet lily that grows up to 8' tall. It produces clusters of large, open-faced, pinkish purple flowers with creamy margins and dark speckles.

L. 'Stargazer' is a vigorous Oriental lily that grows 3–5' tall. It bears clusters of vivid pink flowers with white margins and tips and darker pink speckles.

L. superbum (American turk's cap lily, swamp lily) is a species lily with tiered whorls of leaves spaced along the stems. It grows 5–10' tall. Pyramidal flowerheads bear 30–40 orange-red, darker spotted pendant flowers in July and August. It grows best in moist soil in full sun or partial shade.

Problems & Pests

These plants are quite resistant to most diseases and insect pests with the exception of slugs, which may feast on the young plants. More troublesome can be mammal pests like squirrels, voles, rabbits and deer, all of which consider all parts of lilies to be delectable snacks.

L. martagon 'Northern Explorer' (above)

L. martagon var. *dalmaticum* (center), *L.* hybrid (below)

Lily-of-the-Valley

Convallaria

Height: 6–12" **Spread:** indefinite **Flower color:** white, pink
Blooms: spring **Zones:** 2–7

GARDEN WRITER ALLAN ARMITAGE SUGGESTS THAT FATHERS OF daughters should plant a large patch of lily-of-the-valley and talk the bride-to-be and her fiancé into a spring wedding to avoid the exorbitant cost of having to buy these highly sought after traditional bridal flowers during the wedding season. They wouldn't have to plant too far in advance, either, as lily-of-the-valley is a quick-spreading groundcover. The delicacy of the tiny, scalloped bells dangling from 6" stems hiding demurely amongst its leaves belies the potency of the intoxicating perfume that floats on the breeze wherever they bloom. *Muguet de bois*, as it is called in France, is featured by florists every year on May Day, as it is considered a *porte-bonheur*—a "carrier of happiness." Commuters everywhere mill about, clasping their lily-of-the-valley bouquets, to be offered to girl- or boyfriend, husband or wife, boss, dinner host, friend, etc. Even the Metro is muguet-scented.

Planting

Seeding: Not recommended; easy to propagate by division

Planting out: Spring or fall

Spacing: About 12" (plants spread quickly to fill an area)

Growing

Lily-of-the-valley grows well in any light from **full sun to full shade**. The soil should ideally be of **average fertility, humus rich and moist**, but almost any soil conditions are tolerated. This plant is drought resistant.

Division is rarely required but can be done whenever you need plants for another area or to donate to someone else's garden. The pairs of leaves grow from small pips, or eyes, that form along the spreading rhizome. Divide a length of root into pieces, leaving at least one pip on each piece.

Tips

This versatile groundcover can be grown in a variety of locations. Naturalize it in a woodland garden, perhaps bordering a pathway or beneath shade trees where little else will grow. It is also a good groundcover in a shrub border, where its dense growth and shallow roots will keep the weeds down but won't interfere with the shrubs' roots.

Lily-of-the-valley can be quite invasive. Avoid growing it with plants that are less vigorous and likely to be overwhelmed, such as alpine plants in a rock garden. Give lily-of-the-valley plenty of space to grow and let it go. Avoid planting it where you may later spend all your time trying to get rid of it.

C. majalis var. *rosea*

This plant is well known for the delightful fragrance of its flowers. In fall, dig up a few roots and plant them in pots. Keep the pots in a sheltered spot outdoors, such as a window well, or in a cold frame or unheated porch, for the winter. Check the pot periodically to make sure it hasn't dried out completely; if it has, water enough to moisten the soil. In early spring, bring the pots indoors. The plants will sprout and flower early, and you can enjoy the delicious scent.

Recommended

C. majalis forms a mat of foliage. It grows 6–10" tall and spreads indefinitely. In spring it produces small, arching stems lined with fragrant, white, bell-shaped flowers. **Var.** *rosea* ('Rosea') has light pink or pink-veined flowers; it is a less vigorous grower than the species.

Problems & Pests

Rare problems with molds and stem rot can occur.

Lilyturf
Liriope

Height: 8–18" **Spread:** 18" **Flower color:** shades of purple and blue **Blooms:** late summer through mid-fall **Zones:** 6–9

LILYTURF IS A POPULAR GROUNDCOVER, ALMOST IMPERVIOUS TO drought, heat, humidity and most garden pests and diseases. The plantings of *L. muscari* 'Variegata' at my house were here when I came, well over 20 years ago, and are still performing well. I only divided them once, and that was simply because I was moving them from one spot to another. The only care lilyturf requires is having a crewcut each spring to get rid of its scruffy, dry winter "do" to make room for the new growth.

Planting
Seeding: Start seed outdoors in spring

Planting out: Spring

Spacing: 12–18"

Growing

Lilyturf grows best in **light** or **partial shade** and tolerates both **full sun and full shade** well. The soil should be of **average fertility, humus rich, acidic, moist and well drained**. A sheltered location is preferable because the evergreen leaves are prone to drying out in winter. Divide clumps in spring.

Tips

Lilyturf makes a fantastic, dense groundcover, ideal for keeping weeds down in a variety of locations. Include it in beds, borders, woodland gardens, near water features or under trees where grass won't grow.

An electric handclipper or weed-whacker can make the task of shearing off spent foliage in late winter less of a chore. Be sure to do this before new growth emerges. For large expanses, use a lawnmower with a bagger. No cleanup!

Recommended

L. muscari forms a mass of low clumps of strap-shaped, evergreen leaves. It grows about 8–18" tall and spreads about 18". It bears spikes of purple flowers from late summer through fall. **'Big Blue'** bears large spikes or purple-blue flowers. **'Monroe White'** bears white flowers. **'Pee Dee Gold Ingot'** has golden yellow to chartreuse leaves that mature from a bright yellow. The flowers are light purple. **'Variegata'** has green and creamy, white-striped leaves and bears purple flowers.

L. muscari 'Monroe White' (above), *L. muscari* (below)

L. spicata (creeping lilyturf) is a rapidly spreading, potentially invasive species that is not clump-forming, making it useful for lawn replacement and erosion control. **'Silver Dragon,'** excellent for lighting up a dark area, does not grow as densely as most lilyturf. It has slender, highly variegated, green and white leaves and lavender flowers, and stands about 12" tall.

Problems & Pests

Rare problems with root rot, anthracnose and slugs can occur.

Lungwort

Pulmonaria

Height: 8–24" **Spread:** 8–36" **Flower color:** blue, red, pink, white, purple
Blooms: spring **Zones:** 3–8

FOR THIS PLANT'S UNFORTUNATE BOTANICAL AND COMMON
names, blame the Doctrine of Signatures, a centuries-old theory alleging that the
appearance of a plant dictated God's intended use for it. Someone decided this
one's spotted leaves resembled diseased lungs, (now there's a sales pitch for it, if
ever there was one), so the poor thing was ground up and used for treating pul-
monary problems. In actuality, lungworts are beautiful plants that flower with
the first breath of spring. The only relationship to disease that comes to mind is
that if ever the leaves develop powdery mildew, a nonlethal
albeit normally unsightly affliction to which they are prone,
the characteristic white mottling on the leaves camouflages
the problem nicely, and just adds to the plant's virtues.

*It is rumored that the lungs of Philippus Aureolus
Theophrastus Bombastus von Hohenheim (1493–
1541), the Doctrine of Signature's most famous
proponent, breathed their last in exhaustion after
repeatedly trying to introduce himself to a hearing-
impaired patient.*

Planting

Seeding: Start seed in containers outdoors in spring; plants don't consistently come true to type

Planting out: Spring or fall

Spacing: 12–18"

Growing

Lungworts prefer **partial to full shade**. The soil should be **fertile, humus rich, moist and well drained**. Rot can occur in very wet soil.

Divide in early summer after flowering or in fall. Provide the newly planted divisions with lots of water to help them re-establish. Shear plants back lightly after flowering to deadhead and show off the fabulous foliage and to keep the plants tidy.

Tips

Lungworts make useful and attractive groundcovers for shady borders, woodland gardens and pond and stream edges.

P. longifolia (above), *P.* mixed species (below)

Recommended

P. angustifolia (blue lungwort) forms a mounded clump of foliage. The leaves have no spots. This plant grows 8–12" tall and spreads 18–24". Clusters of bright blue flowers, held above the foliage, are borne from early to late spring. **'Blue Ensign'** bears large, purple-blue flowers. This plant is mildew-resistant.

P. longifolia (long-leaved lungwort) forms a dense clump of long, narrow, white-spotted green leaves. It grows 8–12" tall, spreads 8–24" and bears clusters of blue flowers in spring or even earlier, as the foliage

P. saccharata 'Berries and Cream' (above)
P. longifolia 'Bertram Anderson' (below)

emerges. **Ssp.** *cevennensis* has long, narrow, dark green leaves that are so heavily spotted that they are almost completely silver. **'Bertram Anderson'** has narrow, very spotted leaves and bears clusters of bright blue flowers.

P. officinalis (common lungwort, spotted dog) forms a loose clump of evergreen foliage, spotted with white. It grows 10–12" tall and spreads about 18". The spring flowers open pink and mature to blue. This species was once grown for its reputed medicinal properties, but it is now valued for its ornamental qualities. **'Cambridge Blue'** bears many blue flowers. **'Roy Davidson'** has light blue flowers. **'Sissinghurst White'** has pink buds that open to white flowers. The leaves are heavily spotted with white.

P. rubra (red lungwort) forms a loose clump of unspotted, softly hairy leaves. It grows 12–24" tall and spreads 24–36". Bright red flowers appear in early spring. This species tends to be less drought tolerant than other lungworts and often transfers this trait to its hybrids. **'Redstart'** has pinkish red flowers.

P. saccharata (Bethlehem sage) forms a compact clump of large, white-spotted, evergreen leaves. It grows 12–18" tall, with a spread of about 24". The spring flowers may be purple, red or white. This species has given rise to many cultivars and hybrids. **'Berries and Cream'** has foliage heavily spotted with silver and bears raspberry red flowers. **'Mrs. Moon'** has pink buds that open to a light purple-blue. The leaves are dappled with silvery white spots. **'Pink Dawn'** has dark pink flowers that age to purple.

P. **'Silver Streamers'** has slightly ruffled, almost entirely silver, marked leaves. The flowers are blue. It grows 8–12" tall and spreads 12–18".

P. **'Spilled Milk'** is a compact plant with silvery green splotched leaves and pink flowers.

Problems & Pests

These plants are generally problem free but may become susceptible to powdery mildew if the soil dries out for extended periods. Remove and destroy damaged leaves.

P. saccharata (above), *P. saccharata* 'Mrs. Moon' (below)

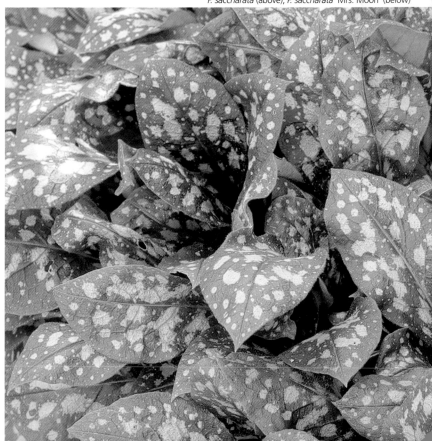

Lupine
Lupinus

Height: 24–36" **Spread:** 12–18" **Flower color:** white, cream, pink, yellow, blue, purple, often bi-colored **Blooms:** spring and early summer **Zones:** 3–8

LUPINES ARE PERSNICKETY, SO WHEN I FOUND A FEW POTS OF Russell hybrids at the supermarket, I figured I'd treat them as annuals. Astonishingly, they gave a long-lasting display, which, by the following year, seeded around, filling 60 square feet with waves of color from spring till frost. The lupine show was a wonder, drawing envious crowds, begging to know the secret of my success. I could have mischievously distributed packets of Magic Lupine Grow, formulated from pulverized "anything" and finely minced Twinkie wrappers, and chortled over their reported results.

Truth is, although lupines supposedly don't like hot, humid weather, mine grew in just that. They don't like lime, but some seeded into the lime-treated lawn. I waved the seed pods over shady places and sunny spots, and they prospered. They grew where soil was hard, arid and infertile or wet and rich, then not at all in similar situations. Try your hand at lupine luck. Who knows?

L. Russell Hybrids

Planting

Seeding: Soak seeds in water for 24 hours, then plant them outdoors in late fall or early spring

Planting out: Spring or fall

Spacing: 12"

Growing

Lupines grow well in **full sun** and **partial shade** in a location sheltered from strong winds. The soil should be **fertile to average, sandy, slightly acidic and well drained**. Plants don't need dividing and resent having their roots disturbed. The small offsets that form at the bases can be transplanted when young as can the seedlings that are bound to sprout up around the plants. Deadheading keeps plants looking tidy but reduces the likelihood of self-seeding.

Tips

Lupines are lovely when massed together in beds and borders, cottage-style gardens and when naturalized in meadow plantings.

Recommended

L. **Russell Hybrids** were developed from the cross-breeding of several species of lupine, the result being compact plants that bear flowers in a wide range of colors. They grow 24–36" tall and spread 12–18". The spring and early summer flowers are produced in a wide range of solid and bi-colors, though blue tends to dominate in self-seeded offspring. Occasionally, a stunning luminous purple, pink or other pleasant surprise may appear.

Problems & Pests

Possible problems with aphids, slugs, leaf spots, downy and powdery mildew and rust can occur.

Mayapple

Podophyllum

Height: 6–30" **Spread:** 12–36" **Flower color:** red, purple, white, pink
Blooms: spring, summer **Zones:** 5–8

OUR NATIVE MAYAPPLES MAKE A HANDSOME, VIGOROUS WOODLAND groundcover, but the fancy Asian imports and hybrids make fabulous specimen plants that halt viewers, mouths agape, asking "Whatever *is* that?" All parts of these plants are poisonous—leaves, stems, roots and fruit—except for the **fully ripened** fruit or "apples" of our native species (not sure about the Asian fruits.) This seeming self-defense mechanism protects the fruit until seeds mature. Then birds, mammals and, especially, box turtles eat them, and the chances of the seeds germinating once they've gone through their digestive tracts are purportedly greatly improved. Humans have eaten them, too. Indian apple, ground lemon, raccoon berry and hog apple are some folk names. Despite the toxicity, herbalists and physicians have long used mayapple root extracts for many purposes, and podophyllotoxin, a chemical from the plant, is an anti-cancer drug. Although mayapple is used to treat warts and skin cancers, handling the plants can cause rashes in people with sensitive skin.

Planting

Seeding: Start seed in a cold frame in late summer or fall

Planting out: Spring

Spacing: 12–24"

Growing

Mayapples grow well in **partial** or **full shade**. The soil should be **fertile, humus rich and moist**. Divide in spring or late summer to propagate more plants. Mulch in winter to protect plants from freeze-thaw cycles.

Tips

These plants are naturally at home in a moist woodland garden and make an excellent addition to a pondside planting. Include them in shaded borders and keep the soil mulched to hold in moisture.

Recommended

P. delavayi is a clump-forming plant that grows 6–18" tall with a spread of about 18". The velvety green foliage is mottled with red, bronze and purple. Each leaf has five to eight lobes, each lobe divided into three more lobes. Pinkish red flowers in summer are followed by apricot-colored fruit.

P. hexandrum (Himalayan mayapple) forms a clump of green- and purple-splattered leaves. It grows about 18" tall and spreads about 12". Flowers are brilliant pink, followed by bright red fruit.

P. 'Kaleidoscope' is a dazzling hybrid with 18" hexagonal, umbrella-like, star-shaped leaves that are patterned like a kaleidoscope in green, splashed with silver,

P. peltatum (above & below)

The genus name is derived from anapodophyllum, *or "duck's-foot leaf." Now, aren't you glad they shortened it?*

cream, purple-black and bronze. It bears clusters of up to 20 large, 2" across, wine-red flowers hanging just below the leaf in summer. It grows about 24" tall, with an equal spread.

P. peltatum (mayapple, American mandrake) forms a low, spreading mass of large, glossy umbrellas, sometimes so dense that the earth beneath remains dry even during a torrential rainfall. It grows to a height of about 18" and spreads up to 4'. Fragrant white or light pink flowers are borne in spring. The "apples" start off green and turn yellow.

P. peltatum (above & below)

P. pleianthum (Chinese mayapple) forms an upright clump 18–30" tall with a spread of up to 36". The star-shaped leaves are glossy green and the summer flowers are deep red or purple, followed by silvery fruit.

Problems & Pests

Slugs may attack young foliage.

All parts of these plants are poisonous, except the **fully ripe** *fruit of our native species. Wear gloves when handling plants to avoid possible rash.*

P. hexandrum (above), *P. pleianthum* (below)

Maidenhair Fern

Adiantium

Height: 12–24" **Spread:** 12–24" **Flower color:** grown for foliage
Blooms: doesn't flower **Zones:** 2–8

GRACEFUL, DELICATE AND ONE OUR MOST ELEGANT NATIVE PLANTS, maidenhair fern is nevertheless very hardy and easy to grow. I have had some for years that seem to require no maintenance. They are noted for their ability to thrive among leaf litter under canopy trees, waking up with nary a tress out of place. The genus name means "repels water," and, indeed, raindrops quickly trickle off the fronds to the ground, leaving them nearly dry. Coral bells, trillium, and lungwort make friendly companions.

A. pedatum *is native to the whole Atlantic seaboard of North America, to the center of the continent. In the western half of the continent, its environmental niche is filled by* A. aleuticum, *the western maidenhair.*

A. pedatum (all photos)

Planting

Seeding: Sow spores outdoors in early fall

Planting out: Spring

Spacing: 12–24"

Growing:

Maidenhair fern is not at all sun tolerant. It grows best in **light to partial shade** and tolerates full shade well. It does poorly under droughty conditions. The soil should be of **average fertility, humus rich, moist and well drained.** Maidenhair fern will rarely need dividing. It dies back entirely in winter and division can be done in autumn or spring to propagate more plants.

Tips

Maidenhair fern makes a lovely and delicate addition to woodland gardens and shaded beds and borders. It can also be included in the damp soil near a water feature.

Recommended

A. pedatum forms a slowly spreading mound of delicate, arching fronds. Over a long period, it can become a continuous groundcover. It can be divided to keep it a clumping fern with only a 24" width or less. The light green leaflets stand out against the black stems. The whole plant turns bright yellow in fall. Most other species of maidenhair fern are better suited to greenhouses and as houseplants because they are far too tender to survive the winter in Pennsylvania.

Problems & Pests

Rarely suffers from any problems.

Meadow Rue

Thalictrum

Height: 2–6' **Spread:** 12–36" **Flower color:** pink, purple, yellow, white **Blooms:** summer, fall **Zones:** 3–8

OF OVER 100 MEADOW RUE SPECIES, ONLY A FEW, IF ANY, SEEM TO be offered in our nurseries and garden centers. For some others, you may have to resort to mail-ordering from catalogs or lists. They should be more popular here, as they are deer resistant and, though the foliage of many resemble delicate columbine leaves, they never suffer from leaf miner, which so often plagues the latter. The fine, ferny foliage, blue-green on many forms, and airy puffs of flowers are the key assets. The larvae of many butterfly species rely on meadow rue as a food source. The birds and the bees like them, too. *Thalictrum* tea is supposed to relieve congestion. The seeds of one species are used to extract thalicarpine, a cancer-fighting agent.

Planting

Seeding: Direct sow in fall or start indoors in early spring; soil temperature should be 70° F

Planting out: Spring

Spacing: 12–24"

Growing

Meadow rue prefers **light** or **partial shade** but tolerates full sun with moist soil. The soil should be **humus rich, moist and well drained**. This plant rarely needs to be divided. In fact, meadow rue dislikes being disturbed, and plants may take a while to re-establish once they have been divided. If necessary for propagation, divide in spring as the foliage begins to develop.

T. rochebruneanum 'Lavender Mist' (above)
T. aquilegifolium (below)

Tips

In the middle or at the back of a border, meadow rue makes a soft backdrop for bolder plants and flowers and is beautiful when naturalized in an open woodland or meadow garden.

These plants often do not emerge until quite late in spring. Mark the location where they are planted so that you do not inadvertently disturb the roots if you are cultivating their bed before they begin to grow.

Do not place individual plants too close together because their airy stems can become tangled. Consider giving the taller meadow rues some support if they are in an exposed location because a good wind may topple them.

Recommended

T. aquilegifolium (columbine meadow rue) forms an upright mound 24–36" tall, with an equal spread. Fluffy purple flowers are borne in early summer. The leaves are similar in appearance to those of columbines. **'Thundercloud'** ('Purple Cloud') has dark purple flowers. **'White Cloud'** has white flowers.

T. **'Black Stockings** is a long-legged beauty that flaunts its height with ebony stems, ferny foliage and airy lavender flower clusters in late spring or early summer. Growing about 6' tall, it is a distinguished presence at the back of a border or edge of a woodland.

T. delavayi (Yunnan meadow rue) forms a clump of narrow stems that usually need staking. It grows 4–5'

T. delavayi 'Hewitt's Double' (above & below)

tall and spreads about 24". The foliage is more fern-like than *T. aquilegifolium*, and is reputed to be deer resistant. It bears fluffy purple or white flowers from mid-summer to fall. **'Hewitt's Double'** is a popular cultivar that produces many tiny, purple, pompom-like flowers.

T. rochebruneanum (lavender mist meadow rue) forms a narrow, upright clump 3–6' tall and 12–24" wide. The late-summer blooms are lavender purple and have numerous distinctive yellow stamens. Plants are self-supporting (like you wish your children were). They are less likely to need staking than other species.

Problems & Pests

Infrequent problems with powdery mildew, rust, smut and leaf spot can occur.

T. aquilegifolium (above)
T. rochebruneanum 'Lavender Mist' (below)

Meadow rue flowers are generally petal-less. Their unique blooms consist of showy sepals and stamens.

Meadowsweet

Filipendula

Also called: Queen-of-the-Prairie
Height: 2–8' **Spread:** 18"–4' **Flower color:** white, cream, pink, red
Blooms: late spring, summer, fall **Zones:** 3–8

MEADOWSWEET CONTAINS SALICYLIC ACID AND OTHER ingredients used to make aspirin, and was, for centuries, a favorite analgesic and herbal remedy for a variety of purposes. In fact, the "spir" in aspirin comes from *Spirea ulmaria*, as *F. ulmaria* was known. Indeed the fluffy, fuzzy flowers resemble those of spireas. Oil from the buds was used in perfume and flowers were soaked in rainwater as complexion water. The flowers were also used as paint brushes. Such a useful plant also looks great in a border or bog and alongside a pond with other moisture lovers at its feet. It is tall, handsome and fragrant, and offers a number of color options for bloomtime.

Planting

Seeding: Germination can be erratic; start seed in cold frame in fall and keep soil evenly moist

Planting out: Spring

Spacing: 18–36"

Growing

Meadowsweets prefer **partial** or **light shade**. Full sun is tolerated if the soil remains sufficiently moist. The soil should be **fertile, deep, humus rich and moist**, except in the case of *F. vulgaris*, which prefers dry soil.

Divide in spring or fall. You may need a sharp knife to divide these plants because they grow thick, tough roots. Meadowsweet tends to self-seed, and if dividing this perennial seems daunting, transplanting the seedlings could be an easier way to get new plants. These plants can be deadheaded if desired, but the faded seedheads are quite attractive when left in place.

Tips

Most meadowsweets are excellent plants for bog gardens or wet sites. Grow them alongside streams or in moist meadows. Meadowsweets may also be grown in the back of a border, as long as they are kept well watered. Grow *F. vulgaris* if you can't provide the moisture the other species need.

The flowers of *F. ulmaria*, once used to flavor mead and ale, are now becoming popular as a flavoring for vinegars and jams. They may also be made into a pleasant wine, which is made in much the same way as dandelion wine.

F. ulmaria cultivar (above), *F. rubra* (below)

Recommended

F. purpurea (Japanese meadow-sweet) forms a clump of stems and large, deeply lobed foliage. It grows up to 4' tall and spreads about 24". Pinkish red flowers fade to pink in late summer. **'Elegans'** has fragrant white flowers. The spent flowerheads develop an attractive red tint.

F. rubra (queen-of-the-prairie) forms a large, spreading clump 6–8' tall and 4' in spread. It produces clusters of fragrant pink flowers from early to mid-summer. **'Venusta'** bears very showy pink flowers that fade to light pink in fall.

F. ulmaria (queen-of-the-meadow) grows 24–36" tall and 24" wide. It bears cream white flowers in large clusters. **'Aurea'** has yellow foliage

F. ulmaria (above & below)

that matures to light green as the summer progresses. **'Flore Pleno'** has double flowers.

F. vulgaris (dropwort, meadowsweet) is a low-growing species up to 24" tall and 18" wide. **'Flore Pleno'** has white double flowers. **'Rosea'** has pink flowers.

Problems & Pests
Powdery mildew, rust and leaf spot can be troublesome.

The flowers of F. ulmaria *were often used to flavor ales and mead in medieval times, giving rise to the name meadowsweet, from the Anglo-Saxon* medesweete.

F. ulmaria cultivar (above), *F. ulmaria* (below)

Mondo Grass

Ophiopogon

Height: 4–12" **Spread:** 8–12" **Flower color:** white, pale purple **Blooms:** summer **Zones:** 6–10

OPHIOPOGON—THE NAME MEANS "SNAKE'S BEARD" AND YOUR guess is as good as mine as to why—is a fantastic boon to modern garden design. Picture these low and linear babies, especially the miniature ones and the black kinds, landscaping a wooden deck or poolside with clean lines and spare simplicity. Perhaps try edging a walk with them. How about in a rock garden? Maybe plant them among your spring bulbs. Mondo grass is also great for pots or troughs. Think of the interesting combinations you can make by integrating a low-growing, spidery or spiky, dark plant with any of your tall or fluffy favorites. Plant black mondo grass with contrasting colors, such as yellow-leafed hostas and delicate ferns, silver-leaved plants or—face it, there's not much that doesn't look good with basic black.

Planting

Seeding: Start seed in a cold frame in fall

Planting out: Spring or fall

Spacing: 8–12"

Growing

Mondo grass grows well in **full sun** or **partial shade**. The soil should be **fertile, humus rich, slightly acidic, moist and well drained**. Divide plants in spring, just as they begin to sprout new leaves. These plants may also be grown in containers.

Tips

Mondo grass makes an excellent groundcover and edging plant for beds and borders. Black mondo grass is an excellent contrast plant, creating stand-out displays with yellow-, blue- or silver-leaved plants. If you've tried mondo grass and it has died off over the winter, try a mulch of evergreen branches to provide it with winter protection without smothering it, which causes rot.

Recommended

O. japonicus forms low, grass-like clumps of dark green foliage. It bears clusters of white flowers in summer. It grows 8–12" tall and spreads about 12". **'Compactus'** is a wee plant at 2" high. **'Kyoto Dwarf'** only grows 4" tall. **'Nanus'** (dwarf mondo grass) grows half the size of the species, 4–8" tall with a spread of about 8".

O. planiscapus forms a compact clump of strap-shaped, dark green leaves. It grows 4–8" tall with a spread of about 12". It bears pale purple flowers in summer. **'Ebony**

O. planiscapus 'Nigrescens' (above), *O. japonicus* (below)

Knight,' **'Black Night,'** and **'Ebony Night'** are other available 4–6" tall cultivars, possibly all the same plant. **'Nigrescens'** ('Black Dragon,' 'Nigra'; black mondo grass) has striking dark, almost black, foliage.

Problems & Pests

Slugs may feed on new leaves in spring.

Monkshood

Aconitum

Height: 3–6' **Spread:** 12–18" **Flower color:** purple, blue, white
Blooms: mid- to late summer **Zones:** 3–8

MONKSHOOD'S FLORAL SPIKES AND DEEPLY CUT FOLIAGE LOOK
somewhat like delphinium's, but these plants are far easier to grow.
They bloom from late summer to fall, their flower spikes gently
swaying and leaning in the woodland or mixed border, their helmet-shaped blooms luring bees. Despite its innocent looks,
monkshood has, possibly, the most sinister reputation among
plants. All its parts are toxic. In mythology, monkshood was
Medea's poison of choice. Athena sprinkled it on the head
of the weaver Arachne to turn her into a spider. In liter-
ature, Shakespeare's Laertes, when facing Hamlet
in a duel, covered his blade in juices from
monkshood, called wolfbane in those days. Its
theatrical fame resurfaces in the movie *Wolf
Man* when Lon Chaney, Jr. is informed he
will become a werewolf "on the night
of the full moon when the wolfbane
blooms." Roman Emperor
Claudius I was slain by his
own physician, who slipped
him monkshood. In fact, it
was so prevalent in political
assassinations that its cultivation was banned,
under penalty of—
what else?—death.

Planting

Seeding: Germination may be irregular. Seeds direct sown in spring may bloom the following summer; seeds sown later will not likely bloom until the third year.

Planting out: Spring; bare-rooted tubers may be planted in fall

Spacing: 18"

Growing

Monkshoods grow well in **full sun** and tolerate light or partial shade. These plants will grow in any moist soil, but they prefer to be in a **rich soil with lots of organic matter worked in**.

A. napellus (above), *A. cammarum* 'Bicolor' (below)

Monkshoods prefer not to be divided, as they may be slow to re-establish. If division is desired to increase the number of plants, it should be done in late fall after blooming or in early spring. When dividing or transplanting monkshoods, the crown of the plant should never be planted at a depth lower than at which it was previously growing. Burying the crown any deeper will cause it to rot and the plant to die.

Tall monkshoods may need to be staked. Peony hoops or tomato cages inserted around young plants will be hidden as the plants fill in.

The upper petals of monkshood flowers are fused, making an enclosure that looks like the cowl worn by medieval monks.

A. napellus (above), *A. x cammarum* 'Bicolor' (below)

Tips

Monkshood plants are perfect for cool, boggy locations along streams or next to ponds. They make tall, elegant additions to woodland gardens in combination with lower-growing plants. Do not plant monkshoods near tree roots because these plants cannot compete with trees.

Monkshoods, like their cousins the delphiniums, prefer conditions on the cool side. They will do poorly when the weather gets hot, particularly if conditions do not cool down at night. Mulch the roots to keep them cool; keep plants well watered; and trim back faded foliage in summer to encourage new growth to fill in when cooler fall weather arrives. Be careful not to cut back too hard or the plant will fail to flower.

Recommended

A. 'Bressingham Spire' bears dark purple-blue flowers on strong spikes. It grows up to 36" tall but needs no staking.

A. x *cammarum* (Cammarum hybrids) is a group that contains several of the more popular hybrid cultivars. **'Bicolor'** (bicolor monkshood) has blue and white, helmet-shaped flowers. The flower spikes are often branched.

Always take care to avoid getting the poisonous juice from these plants in open wounds or in your mouth.

A. carmichaelii (azure monkshood) forms a low mound of basal leaves from which the flower spikes emerge. The foliage generally grows to about 24" in height, but the plant can grow 6' tall when in flower. Purple or blue flowers are borne a week or so later than those of other species. **'Arendsii'** produces dark blue flowers on strong spikes that need no staking.

A. napellus (common monkshood) is an erect plant that forms a basal mound of finely divided foliage. It grows 3–5' tall, spreads 12–18" and produces dark purple-blue flowers. It yields the drug *aconite*, which serves as a cardiac sedative and an analgesic in liniments. It is very dangerous and should never be self-administered.

Problems & Pests

Problems with aphids, root rot, stem rot, powdery mildew, downy mildew, wilt and rust can occur.

Aconitum *may come from the Greek* akoniton, *meaning "dart." The ancient Chinese and the Arabs used the juice of monkshood to poison arrow tips. Europeans used it to poison wolves and an occasional person or two.*

Oriental Poppy

Papaver

Height: 18"–4' **Spread:** 18–36" **Flower color:** red, orange, pink, white; often with black blotches in center of flower **Blooms:** spring, early summer **Zones:** 3–7

THE ORIENTAL POPPY BLOOMS FOR ONLY A FEW WEEKS AT BEST, but what a splendid show it puts on for that time. Furthermore, it offers encore performances for decades, as it is a very long-lived perennial. The colors are flashy, perhaps too garish for some gardeners, but every year new cultivars appear, widening and softening the range. Silky 'Patty's Plum' and some of the New York series cultivars in sherbet shades will please most critics, and ruffled white, peach-rimmed 'Picotée' is a demure beauty that should fit in many places. When poppy foliage yellows and fades, and before it reappears in fall, camouflage the vacant spot left behind by planting annuals or daylilies, grasses, Siberian iris, phlox, boltonia or other perennials nearby.

Planting

Seeding: Direct sow in spring or fall

Planting out: Spring; poppies like to be planted deeper than most perennials. Cover the crown with 3" of soil.

Spacing: 24"

Growing

Grow Oriental poppy in **full sun**. The soil should be **average to fertile and must be well drained**.

Plants die back after flowering and send up fresh new growth in late summer. This growth should be left in place for winter insulation. Division is rarely needed but may be done in fall once new rosettes begin to form. Oriental poppy can also be propagated by root cuttings taken from dormant plants in mid- to late summer.

Tips

Small groups of Oriental poppy look attractive in an early summer border, although they may leave a bare spot during the dormant period in summer. Baby's breath and catmint make good companions and will fill in any blank spaces.

Recommended

P. orientale forms an upright, oval clump 18"–4' tall and 24–36" wide. Red, scarlet, pink or white flowers with prominent black stamens are borne in late spring and early summer. There are numerous cultivars. **'Allegro'** has bright, scarlet red flowers. **'Carneum'** bears salmon pink flowers early in the season. **'Patty's Plum'** offers a color breakthrough with dusky mulberry blooms,

P. orientale (above & below)

Use of poppy seeds in cooking and baking can be traced as far back as the ancient Egyptians.

P. orientale 'Patty's Plum' (above)

P. orientale 'Picotee' (above)

For cut flowers, seal the cut end of each stem with a flame or boiling water.

though they tend to recede in the border. **'Picotée'** has papery white blossoms with deep salmon edges varying in width. **'Pizzicato'** is a dwarf cultivar, with flowers in a wide range of colors. It forms a mound 18–24" tall, with an equal spread. **'Türkenlouis'** has vivid scarlet flowers with fringed petals and black centers. It grows about 36" tall. The **New York series** of *P. orientale* cultivars includes **'Brooklyn,'** with flowers of rosy pink; **'Central Park,'** with deep wine red flowers; **'Harlem,'** bearing large, burgundy-rose flowers with black petal bases,

reblooms in September and has lon-
ger-lasting foliage; **'Manhattan,'** with
luscious, lavender-pink blossoms;
'Queens,' with delicate pink flowers
with lighter pink margins; **'Soho,'**
with rich, mauvy red flowers; and
'Staten Island,' with purple-red
flowers that rebloom in September
and long-lasting foliage.

Problems & Pests

Problems with powdery mildew, leaf
smut, gray mold, root rot and damping off are possible but rare in well-
drained soil. This plant is deer
resistant.

P. orientale cultivar (above), *P. orientale* 'Türkenlouis' (above)

Ostrich Fern

Matteuccia

Also called: Shuttlecock Fern
Height: 3–5' **Spread:** 12–36" **Flower color:** brown fertile fronds
Blooms: late summer **Zones:** 1–8

OUR LARGEST NATIVE FERN, CLASSIC, GRACEFUL, "MOISTURE-frondly" *Matteuccia struthiopteris*, makes an excellent choice where a bold grouping is desired. Ostrich fern makes a vigorous groundcover if planted in a damp, shady spot. The fronds are reminiscent of ostrich feathers, and the upright form of the plant reminds one of a badminton shuttle-cock, hence the common names. *Matteuccia* fiddleheads, as the newly emerging growth (crosiers) that appears each spring are called, are a favorite edible for some for-aging gardeners. They taste somewhat like asparagus and can be served in many ways, including hot with butter, cheese, lemon or hollan-daise sauce, fresh in sal-ads, cold on a slice of hard-boiled egg on a cracker, in soup, quiche or on pizza. Cutting only three or four fid-dleheads from each crown won't sap vigor from the plant, and new fronds arise to replace cut ones.

Planting

Seeding: Start spores outdoors in fall

Planting out: Spring or fall

Spacing: 12–24"

Growing

Ostrich fern prefers **partial** or **light shade** but tolerates full shade or full sun if the soil is consistently moist. The soil should be **average to fertile, humus rich, neutral to acidic and moist.** Leaves may scorch if the soil is allowed to dry out for extended periods. These ferns spread aggressively. Unwanted plants can be pulled up and composted or given away to friends.

M. struthiopteris (all photos)

Tips

Ostrich fern appreciates a moist woodland garden and is often found growing wild alongside woodland streams and creeks, where it can form a thicket 5' high. Useful in shaded borders, these plants are quick to spread, to the delight of those who enjoy the young fronds as a culinary delicacy.

Recommended

M. struthiopteris (*M. pennsylvanica*) forms a circular cluster of slightly arching, green fronds. Stiff, brown, fertile fronds stick up from the center of the plant in late summer and persist through winter.

Problems & Pests

Rarely suffers from any problems and is deer resistant.

Pachysandra

Pachysandra

Also called: Japanese Spurge

Height: about 8" **Spread:** 12–18", or more **Flower color:** white, inconspicuous; foliage plant **Blooms:** early spring **Zones:** 4–9

POSSIBLY THE MOST OVERUSED GROUNDCOVER IN THE U. S. TODAY is pachysandra. That being said, it is valuable for creating a dense, lush carpet under trees, shrubs, decks, and over barren areas that beg for something attractive to suppress weeds and provide a backdrop for spring bulbs. I've noticed the plant is also ignored by deer, forming a nice serving platter upon which to offer up the hostas I have interplanted with them. Our native pachysandra, *P. procumbens*, has leaves that are far lovelier, larger and bronze green, taking on red hues in autumn. 'Angola' Allegheny spurge has heavily patterned pewter markings on its olive green leaves—a real stunner. Its only drawbacks are that it is not evergreen, disappearing over the winter, and is a slow grower.

Planting

Seeding: Not recommended
Planting out: Spring or fall
Spacing: 12–18" apart.

Growing

Pachysandra prefers **light to full shade** and tolerate partial shade. The soil should be **moist, humus rich and well drained**. Division is not required, but can be done in spring for propagation.

Tips

Pachysandra is a durable ground-cover, useful under trees, along north walls, in shady borders and in woodland gardens.

Most plants are evergreen and generally need little attention. Shearing back any winter-damaged plants in spring will quickly result in a flush of new growth.

Recommended

P. procumbens (Allegheny spurge, Allegheny pachysandra) is a clump-forming species, native to the southeastern United States. It is hardy to

P. terminalis (all photos)

Zone 5. The semi-evergreen leaves emerge a light green, mature to bronze green and turn reddish in the fall. This cultivar can be used as an accent plant, as a ground cover or in a naturalized, woodland setting. **'Angola'** has olive green leaves sporting silver splotches.

P. terminalis (Japanese spurge) forms a low mass of foliage rosettes. It grows about 8" tall and can spread almost indefinitely. **'Green Sheen'** has smaller, shiny, dark green leaves. This slow spreader is best used on a small scale to provide contrast. **'Variegata'** has white margins or silver-mottled foliage. It is not as vigorous as the species.

Problems & Pests

Problems with leaf blight, root rot and scale insects can occur.

Painted Fern/ Lady Fern

Athyrium

Height: 12–36" **Spread:** 12–36" **Flower color:** no flowers, grown for foliage
Blooms: none **Zones:** 4–8

WHY ATHYRIUMS, AS A GROUP, ARE CALLED "LADY FERNS" PUZZLES
even fern expert John Mickel. Some contend "its extreme variability is like a
woman's changing mind," he says, but that he has "been advised not to
pursue this" theory. Dependable and hardy to Zone 5, most have
fragile fronds that are easily broken, but unless you're planning
to go bumbling through your fern bed, don't hesitate to try
some of the many forms this delicate maiden takes. I have
grown *A. filix-femina* in a relatively sunny spot with
no audible (or visible) complaints from
Mademoiselle; I've moved and messed
with painted ferns, and they persist
unchipped through all their
traumas; and 'The Ghost,' all
silver and lace, grows in a
trough against a north-facing
wall and is as content as,
and more beautiful
than, any fern I've
ever grown.

Planting

Seeding: Start spores outdoors in fall

Planting out: Spring or fall

Spacing: 12–18"

Growing

Painted and lady ferns grow well in **full shade, partial shade** and **light shade.** The soil should be of **average fertility, humus rich, acidic and moist.** Division is rarely required but can be done to propagate desirable plants.

Tips

Painted and lady ferns form an attractive mass of foliage, but they don't spread uncontrollably as some ferns tend to. Include them in shaded borders and moist woodland gardens.

Recommended

A. filix-femina (lady fern) forms a dense clump of lacy fronds. It grows 12–36" tall and spreads about 36". Many cultivars and variations are available, usually with unusual feathery, crested or bunched fronds. **Var. *angustatum* 'Lady in Red'** has lacy green foliage that contrasts with the bright red stems.

A. 'Ghost' was developed from a cross between lady fern and painted fern. It has silvery foliage and a vigorous, upright habit.

A. niponicum **var. *pictum*** (painted fern, Japanese painted fern) forms a clump of dark green fronds with a silvery or reddish metallic sheen. It grows 12–24" tall and spreads 12–18". Cultivars with lovely variations in foliage coloring are available.

A. niponicum var. pictum (above), *A. filix-femina* (below)

Problems & Pests

Occasional problems with rust can occur.

These are quite possibly the most beautiful and well-behaved ferns available.

Pasqueflower

Pulsatilla

Height: 4–12" **Spread:** 8–12" **Flower color:** purple, blue, red, white
Blooms: early to mid-spring **Zones:** 3–7

MOST ROCK GARDENERS KNOW PASQUEFLOWER WELL. IT'S AN
early spring riser with lovely, anemone-like flowers that leave behind beautiful seedheads, rivaling those of clematis for interest. Grow it at border's edge
or in your own rockery, along with early-blooming bulbs and small-scale
alpine plants. Although this buttercup relative is dangerously toxic, it has
nevertheless been used medicinally from ancient to modern times. It has
served as a pain killer; to treat infected sores, varicose veins, venereal disease;
as a diuretic, antispasmotic, sedative, expectorant; and for eye complaints,
toothache, earache—well, you get the picture. Elizabethans dried and powdered the flowers into snuff for headaches. It has become rare in the wild
because of herb harvesting. A green dye is obtained from the flowers.

Planting

Seeding: Sow seed as soon as it is ripe (mid-summer to fall)

Planting out: Spring

Spacing: 8–12"

Growing

Pasqueflower grows well in **full sun** or **partial shade**. The soil should be **fertile and very well drained**. Poorly drained, wet soil can quickly kill this plant. Pasqueflower resents being disturbed in any way. Plant it when it is very young, and don't divide it.

Propagate pasqueflower by carefully taking root cuttings in early spring. You may have to soak the soil around the plant to loosen it enough to get at the roots. Dig carefully to expose a root, then remove it and replant it. Take a look at the 'Propagating Perennials' section in the Introduction for more information about starting root cuttings. Be sure to protect the remaining plant from spring frosts with some mulch if you have taken cuttings from it.

Tips

Grow pasqueflower in rock gardens, woodland gardens and borders and on gravelly banks. It also works well in pots and planters but should be moved to a sheltered location for the winter. An unheated garage or porch will protect from freeze-thaw cycles and excessive moisture. Make sure the pots get some light once the plants begin to show signs of growth.

Pasqueflower is harmful if eaten, and repeated handling may cause skin irritation.

P. vulgaris (all photos)

Recommended

P. vulgaris *(Anemone pulsatilla)* forms a mounded clump of lacy foliage. Flowers in shades of blue, purple or occasionally white are borne in early spring, before the foliage emerges. The seedheads are very fluffy and provide interest when the flowers are gone. **'Alba'** has white flowers. **Var.** *rubra* has bright purple-red flowers.

Problems & Pests

Pasqueflower is rarely troubled by pests or diseases.

Penstemon

Penstemon

Also called: Beardtongue

Height: 9"–5' **Spread:** 6–24" **Flower color:** white, pink, purple, red, blue **Blooms:** late spring, summer, fall **Zones:** 4–8

I MUST ADMIT THAT, EXCEPT FOR THE STALWART SOLDIER 'HUSKER Red,' I haven't had much success keeping penstemons going for too many years, but since others around me apparently have, I will chalk it up to my ineptitude. Perhaps I've tried the wrong ones, or had them sited improperly. Some are short-lived and some are only suited to rock garden conditions, while others are appropriate in a flower bed and would suffer in a dry rock garden. Penstemons listed here have been recommended to me by savvy Pennsylvanians who have been successful growing them in their gardens. Penstemons are beautiful flowers with great variety of height, color, and garden utility, and I encourage people to try them, even if they are a little tricky.

Planting

Seeding: Start indoors in late summer or early spring; soil temperature should be 55°–64° F

Planting out: Spring or fall

Spacing: 12–24"

Growing

Penstemons prefer **full sun** but **tolerate some shade**. The soil should be of **average to rich fertility, sandy and well drained**. These plants are drought tolerant and will rot in wet soil. Mulch in winter with chicken grit or pea gravel to protect the crowns from excessive moisture, especially the smaller species that are from the mountain regions of North America.

P. barbatus cultivar (below)

Divide every two or three years in spring. Pinch plants when they are 12" tall to encourage bushy growth.

Tips

Use penstemons in a mixed or herbaceous border, a cottage garden or a rock garden. These plants are also good for hummingbird gardens. *P. digitalis* 'Husker Red' makes an attractive mass planting.

Twiggy branches pushed into the ground around young penstemon plants will support them as they grow.

Penstemons are particular about having well-drained soil, but at the same time they prefer even moisture and a bit of compost in spring and fall.

P. 'Alice Hindley' (above), P. barbatus cultivar (below)

Recommended

P. **'Alice Hindley'** bears pinkish purple flowers with white throats from mid-summer to fall. It grows 24–36" tall and spreads 12–18".

P. **'Apple Blossom'** bears pink-flushed white flowers from late spring to mid-summer. This rounded perennial grows 18–24" tall, with an equal spread.

P. ***barbatus*** (beardlip penstemon) is an upright, rounded perennial. It grows 18–36" tall and spreads 12–18". Red or pink flowers are borne from early summer to early fall. **'Alba'** has white flowers. **'Elfin Pink'** is very reliable and has compact spikes of pink flowers. It grows up to 18" tall. **'Praecox Nanus'** ('Nanus Rondo') is a compact, dwarf plant that grows about half the size of the species. It bears pink, purple or red flowers. **'Prairie Dusk'** has tall spikes of tubular, rose purple flowers; it blooms over a long season. *P. barbatus* and its hybrids and cultivars are the cold-hardiest (to Zone 3) of the penstemons, and drought tolerant once established.

P. ***digitalis*** (foxglove penstemon, talus slope penstemon) is a very hardy, upright, semi-evergreen perennial. It grows 2–5' tall and spreads 18–24". It bears white flowers, often veined with purple, all summer. **'Husker Red'** combines white flowers with vibrant burgundy foliage that adds season-long interest. Chosen as the 1996 Perennial Plant of the Year by the American Perennial Plant Association, this cultivar was developed for hardiness as well as good looks.

***P.* 'Schoenholzeri'** ('Firebird') has narrow leaves and grows up to 30" tall with a spread of about 24". It bears large scarlet flowers from mid-summer through fall. It likes a somewhat dry location to discourage root rot, and prefers not to be fertilized.

***P.* 'White Bedder'** grows 24–30" tall and 18" wide. Its white flowers become pink tinged as they mature. It blooms from mid-summer to mid-autumn. This hybrid is intolerant of dry shade.

Problems & Pests
Powdery mildew, rust and leaf spot can occur but are rarely serious.

Over 200 species of Penstemon are native to varied habitats from mountains to open plains throughout North and South America.

P. digitalis 'Husker Red' (above & below)

Peony

Paeonia

Height: 24–32" **Spread:** 24–32" **Flower color:** white, yellow, pink, red
Blooms: spring, early summer **Zones:** 2–7

FROM RICKETY OLD FARMHOUSE GARDENS TO POSH ESTATES, peonies have been a perennial favorite since they were brought here in colonial times. Their flamboyance, fragrance, ease of culture and sheer glamour render them irresistible. Peonies are slow growers and often take a couple of years to begin blooming. Once established, they may well outlive their owners. Despite their exotic appearance, these tough perennials can survive winter temperatures as low as -40˚ F. Most herbaceous peonies available are hybrids, but don't ignore the dainty woodland species peonies. I've been growing *P. obovata* along with epimedium, Solomon's seal, Jack-in-the-pulpit, monkshood and other babes in the wood for many years now, and they never fail to delight me when they spring up. Even though the bloom itself is so ephemeral, the seed pods are equally ornamental.

Planting

Seeding: Not recommended; seeds may take two to three years to germinate and many more years to grow to flowering size

Planting out: Fall

Spacing: 24–36"

Growing

Hybrid peonies prefer **full sun** but tolerate some shade. Woodland peonies generally prefer **shade.** The planting site should be well prepared before the plants are introduced. Peonies like **fertile, humus-rich, moist, well-drained soil,** to which lots of compost has been added. Mulch peonies lightly with compost in spring. Too much fertilizer, particularly nitrogen, causes floppy growth and retards blooming. Planting, transplanting and dividing peonies are best done in early fall but may be done in spring as soon as soils are workable. Peonies rarely need dividing, but it is a good way to propagate desirable plants.

Cut back the flowers after blooming and remove any blackened leaves to prevent the spread of gray mold. Red peonies are more susceptible to disease.

Whether you choose to clean most of your perennial garden in fall or spring, it is essential to deal with peonies in fall. To reduce the possibility of disease, clean up and discard or destroy all leaf litter before the snow falls.

P. lactiflora cultivar (above & below)

Tips

These are wonderful plants that look great in a border when combined with other early bloomers. They may be underplanted with bulbs and other plants that will die down by mid-summer, when the emerging foliage of peonies will hide the dying foliage of spring plants. Avoid planting peonies under trees where they will have to compete for moisture and nutrients.

Planting depth is a very important factor in determining whether or not a peony will flower. Tubers planted too shallowly or, more commonly too deeply, will not flower. The buds or eyes on the tuber should be 1–2" below the soil surface.

Place wire tomato or peony cages around the plants in early spring to support heavy flowers. The cage will be hidden by the foliage as it grows up into the wires.

P. lactiflora 'Shimmering Velvet' (above), *P. lactiflora* cultivar (below)

Recommended

Peonies may be listed as cultivars of a certain species or as interspecies hybrids. Hundreds are available.

P. japonica (woodland peony) forms a large clump of upright stems and large, gray-green, lobed leaves with hairy undersides. It grows 18–24" tall, with an equal spread. It bears white 3" wide bowl-shaped flowers in early May followed by pods that split to reveal metallic blue seeds on red stalks. This species makes a great specimen in a woodland garden or scattered throughout a shady perennial border in light to medium shade. It is often mistaken for *P. obovata*, a somewhat similar, equally wonderful plant.

P. lactiflora (Chinese peony) and ***P. officinalis*** (the common garden

P. lactiflora cultivar (above), *P. tenuifolia* (below)

peony) have been hybridized and their offspring generally referred to as "herbaceous peonies." These hybrids are often sold as cultivars. They form clumps of red-tinged stems and dark green foliage, and grow up to 30" tall with an equal spread. They bear single, double and semi-double, fragrant, white, red, rose, fuchsia and pink flowers with yellow stamens. Rarely, a yellow-flowering hybrid may be found. Some popular hybrid cultivars are **'Dawn Pink'** with single, rose pink flowers; **'Duchess de Nemours'** (introduced from France in 1856) with fragrant, white, double flowers tinged yellow at the bases of the inner petals; and **'Sara Bernhardt'** with large, fragrant, pink, double flowers. Most of these require at least six hours of full sun to bloom well, and most bloom in late May.

P. lactiflora cultivar (above, below & opposite page)

P. obovata (woodland peony) forms a large clump of gray-green leaves with slightly hairy undersides. It grows about 24" tall, with an equal spread. Gray stems bear 3" pink or rose, cup-shaped flowers that mature to showy red seedpods with jet black seeds. **Var. *alba*** is a white form. Because it is a woodland species, plant this cultivar in dappled shade close to the front of a shade border.

P. tenuifolia (fernleaf peony) forms an attractive mound of deeply divided, fern-like foliage. It grows 12–18" tall and spreads 18–24". The single red flowers with their bright yellow stamens stand out sharply against the foliage. **'Plena'** bears red double flowers.

Problems & Pests

Peonies may have trouble with *Verticillium* wilt, ringspot virus, tip blight, stem rot, gray mold, leaf blotch and nematodes. Ants enjoy crawling around the unopened buds of the garden peonies, but they do no actual harm.

Peonies can last in gardens for hundreds of years. Because of this, the Philadelphia Centennial Exposition of 1876 used the peony to symbolize the American spirit, ambition and determination to adapt and thrive. To the Chinese and Japanese, the peony is a symbol of prosperity.

Phlox

Phlox

Height: 2"–4' **Spread:** 12–36" **Flower color:** white, orange, red, blue, purple, pink **Blooms:** spring, summer, fall **Zones:** 3–8

ALTHOUGH GARDEN PHLOX, *P. PANICULATA*, IS MOST FAMILIAR, there are more than 60 species of this plant, all but one native to North America. The carpeting moss pink (*P. subulata*) with needle-like foliage, was originally sent from Philadelphia to England in 1745 by John Bartram, founder of America's first botanical garden. Shade-loving *P. divaricata*, woodland phlox, has beautiful, icy blue-violet or pure white flowers. Still more shade loving is *P. stolonifera*, but the majority prefer full sun. Most garden phlox have exceptionally sweet-scented flowers and attract hordes of butterflies. Unfortunately, many older cultivars are susceptible to powdery mildew. If yours succumb, give them plenty of air circulation, try to water only the roots, treat them with fungicide, plant only the newer mildew-resistant forms, or tell people you are planting a white garden like the famous one writer Vita Sackville-West planted. Phlox drop seed promiscuously and, regardless of the color you originally planted, many of the offspring will be magenta (a perfectly nice color.)

Planting

Seeding: Not recommended

Planting out: Spring

Spacing: 12–36"

Growing

P. paniculata and *P. maculata* prefer **full sun**; *P. subulata* prefers **full sun to partial shade**; *P. divaricata* prefers **light shade**; and *P. stolonifera* prefers **light to partial shade** but tolerates heavy shade. All like **fertile, humus-rich, moist, well-drained soil**. *P. paniculata* and *P. maculata* bloom better if fertilized and watered regularly. Divide in fall or spring.

P. stolonifera spreads out horizontally as it grows. The stems grow roots where they touch the ground, and new plants are easily created by detaching the rooted stems in spring or early fall. Do not prune this phlox in fall—it is an early-season bloomer and will have already formed next spring's flowers.

Tips

Low-growing species are useful in a rock garden or at the front of a border. Taller phloxes may be used in the middle of a border and are particularly effective if planted in groups.

To prevent mildew, make sure *P. paniculata* has good air circulation. Thin out large stands to help keep the air flowing. *P. maculata* is more resistant to powdery mildew than *P. paniculata*.

P. paniculata 'Catherine' (above), *P. maculata* (below)

Phlox comes in many forms, from low-growing creepers to tall, clump-forming uprights. The many species can be found in diverse climates, including dry, exposed mountainsides to moist, sheltered woodlands.

P. paniculata (above & below)

Recommended

P. divaricata (woodland phlox) is a low, bushy plant. It grows 6–12" tall and spreads about 12". It produces lavender blue flowers with darker purple centers in early May. There are also white and red-violet cultivars.

P. maculata (early phlox, garden phlox, wild sweet William) forms an upright clump of hairy stems and narrow leaves that are sometimes spotted with red. It grows 24–36" tall and spreads 18–24". Pink, purple or white flowers are borne in conical clusters in the first half of summer. This species has good resistance to powdery mildew. **'Miss Lingard'** bears white flowers all summer. **'Omega'** bears white flowers with light pink centers. **'Rosalinde'** bears dark pink flowers. These cultivars are taller than the species, usually 30" or more.

P. paniculata (garden phlox, summer phlox) blooms in summer. The many cultivars vary greatly in size, growing 20"– 4' tall and spreading 24–36". Many colors are available, often with contrasting centers. **'Bright Eyes'** bears light pink flowers with deeper pink centers. **'David'** was Perennial Plant of the Year for 2002. It bears white flowers and resists powdery mildew. **'Eva Cullum'** has pink flowers with red centers. **'Starfire'** bears crimson red flowers.

P. stolonifera (creeping phlox) is a low, spreading plant. It grows 4–6" tall, spreads about 12" and bears flowers in shades of purple in spring.

This species was Perennial Plant of the Year in 1990.

P. subulata (moss phlox, moss pinks) is very low growing, only 2–6" tall, with a spread of 20" or more. Its cultivars bloom from late spring to early summer in various colors. The foliage is evergreen. **'Candy Stripe'** bears bicolored pink and white flowers.

Problems & Pests

Occasional problems with powdery mildew, stem canker, rust, leaf spot, leaf miners and caterpillars (many of which should be tolerated as they morph into pretty butterflies) are possible.

The genus name Phlox *is from the Greek word for "flame," referring to the colorful flowers of many species.*

P. subulata (above & below)

Pinks

Dianthus

Also called: Carnations

Height: 2–18" **Spread:** 6–24" **Flower color:** pink, red, white, lilac **Blooms:** spring, summer **Zones:** 3–9

IF EVER THERE WAS A WAY TO ENTICE RABBITS INTO YOUR FLOWER beds, carnations (*D. caryophyllus*) are a sure-fire attractant. For some reason, the rabbits don't seem to bother other pinks in my garden, and I'm not sure why. Maybe they're just too lazy to bend. So, I gave up on carnations and stick to winners like *D. gratianopolitanus* (I know—it's just as hard to spell as pronounce) 'Baths Pink'. This is one of the few that hasn't disappeared over the years. Pinks are ideal rockery plants because they need perfect drainage and like conditions dry and sunny. Their intoxicating fragrance alone makes them garden-worthy. I planted something akin to the feathery 'Rainbow Loveliness White' one spring and it went on to that big compost heap in the sky thanks to excess rain that year, but, oh, that spicy perfume while it lasted!

Planting

Seeding: Not recommended; cultivars do not come true to type from seed

Planting out: Spring

Spacing: 10–20"

Growing

Pinks prefer **full sun** but tolerate some light shade. A **well-drained, neutral** or **alkaline soil** is required. The most important factor in the successful cultivation of pinks is drainage—they hate to stand in water. The native habitat of many species is rocky outcroppings. Mix organic matter and sharp sand or gravel into their area of the flowerbed to encourage good drainage, if needed. Pinks may be difficult to propagate by division. It is often easier to take cuttings in summer, once flowering has finished. Cuttings should be 1–3" long. Strip the lower leaves from the cutting. The cuttings should be kept humid, but be sure to give them some ventilation so that fungal problems do not set in. Roots will begin developing in 7–10 days.

Tips

Pinks make excellent plants for rock gardens and rock walls, and for edging flower borders and walkways. They can also be used in cutting gardens and even as groundcovers.

To prolong blooming, deadhead as the flowers fade, but leave a few flowers in place to go to seed. Pinks self-seed quite easily. Seedlings may differ from the parent plants, often with new and interesting results.

D. gratianopolitanus cultivar (above)
D. plumarius (above)

D. gratianopolitanus (above), D. plumarius (below)

Recommended

D. x *allwoodii* (Allwood pinks) are hybrids that form a compact mound and bear flowers in a wide range of colors. Cultivars generally grow 8–18" tall, with an equal spread. **'Aqua'** has pure white, double flowers. **'Doris'** bears salmon pink, semi-double flowers with darker pink centers. It is popular as a cut flower. **'Laced Romeo'** boasts spice-scented red flowers with cream-margined petals. **'Rainbow Loveliness'** is a mix of exceedingly fragrant, frilly, lace-cut flowers of pink, rose and white. **'Sweet Wivelsfield'** bears fragrant, two-toned flowers in a variety of colors.

D. *deltoides* (maiden pinks) grows 6–12" tall and about 12" wide. The plant forms a mat of foliage and flowers in spring. This is a popular species to use in rock gardens. **'Alba'** has bright green foliage and white flowers. **'Brilliant'** ('Brilliancy,' 'Brilliance') bears dark red flowers. **'Zing Rose'** bears carmine red blooms.

D. 'Essex Witch' forms a low mound of evergreen foliage 5–8" tall with a spread of up to 10". It bears semi-double, fragrant, pink flowers with delicately fringed edges.

D. *gratianopolitanus* (cheddar pinks, clove pinks) usually grows about 6" tall but can grow up to 12" tall and 18–24" wide. This plant is long-lived and forms a very dense mat of evergreen, silver gray foliage with sweet-scented flowers borne in summer. **'Bath's Pink'** bears plentiful, light pink flowers and tolerates warm, humid conditions. It spreads up to 24". **'Firewitch'** forms a compact hummock of deep blue leaves. It grows to 6" tall and 6–10" wide and bears deep magenta, single flowers. This cultivar was named Perennial Plant of the Year for 2006. **'Petite'** is even smaller, growing 2–4" tall, with pink flowers.

D. plumarius (cottage pinks) is noteworthy for its role in the development of many popular cultivars known as garden pinks. They are generally 12–18" tall and 24" wide, although smaller cultivars are available. They all flower in spring and into summer if deadheaded regularly. The flowers can be single, semi-double or fully double and are available in many colors. **'Sonata'** bears fragrant double flowers in many colors all summer. **'Spring Beauty'** bears double flowers in many colors. The petal edges are more strongly frilled than those of the species.

Problems & Pests

Providing good drainage and air circulation will keep most fungal problems away. Occasional problems with slugs, blister beetles, sow bugs and grasshoppers are possible. Rabbits seem to find the flowers tasty.

D. deltoides (above)

Cheddar pink is a rare and protected species in Britain. It was discovered in the 18th century by British botanist Samuel Brewer, and it became as locally famous as Cheddar cheese.

D. gratianopolitanus 'Bath's Pink' (below)

D. x *allwoodii* 'Painted Beauty' (below)

Primrose

Primula

Height: 6–24" **Spread:** 6–18" **Flower color:** red, orange, yellow, pink, purple, blue, white **Blooms:** spring, early summer **Zones:** 3–8

THE NAME "PRIMROSE" COMES FROM "PRIMA ROSA," MEANING "first rose of the year." The early arrival of these perky little spring cheerleaders is welcomed by humans, and by butterfly caterpillars, which feed on the leaves. Ants are attracted by the sticky seeds and aid their dispersal, insuring that before long you'll have a carpet of them. (Primroses, that is, not ants.) The candelabra primroses, *C. japonica* and a good number of other species, bloom in May and love moist stream margins and boggy situations. Hundreds of years ago, *Primula* were grown for medicinal purposes, mostly wishful thinking (stem juice rubbed onto the face to remove spots and freckles, for instance.) Often they were amuletic, employed to ward off witches or attract good fortune, and much primrose promise involved their capacity to lure fairies.

Primrose flowers can be made into wine or candied as an edible decoration. The young leaves of P. veris *are also edible and can be added to salads for a hint of spice.*

Planting

Seeding: Direct sow ripe seeds any time of year; start indoors in early spring or in a cold frame in fall or late winter

Planting out: Spring

Spacing: 6–18"

Growing

Choose a location for these plants with **light** or **partial shade**. For most, the soil should be **moderately fertile, humus rich, moist, well drained and neutral or slightly acidic**. Primroses are not drought resistant and will quickly wilt and fade if not watered regularly. Overgrown clumps should be divided after flowering or in early fall. Pull off yellowing or dried leaves in fall for fresh new growth in spring.

Tips

Primroses work well in many areas of the garden. Try them in a woodland area or under the shade of taller shrubs and perennials in a border or rock garden. Moisture-loving primroses may be grown in a bog garden or in almost any other moist spot.

The species with flowers on tall stems look lovely planted in masses. Those with solitary flowers peeking out from the foliage are interesting dotted throughout the garden.

If your primroses always look faded by mid-summer, you may wish to grow them as annuals and just enjoy them in spring and early summer.

P. veris (above & below)

Recommended

P. japonica (Japanese primrose) grows 12–24" tall and 12–18" wide. It thrives in moist, boggy conditions and does poorly if not provided with enough moisture. It is a candelabra flowering type, meaning that the long flower stem has up to six evenly spaced rings of flowers along its length.

P. kisoana forms a basal rosette of hairy foliage about 8" tall and with a spread of up to 16". It bears clusters of tubular flowers in pink in spring. **Var.** *alba* has white flowers.

P. x *polyanthus* (polyantha primrose, polyantha hybrids) usually grows 8–12" tall, with about an equal spread. The flowers are clustered at the tops of stems of variable height. It is available in a wide range of solid colors or bicolors and is often sold as a flowering, potted plant.

P. japonica (above), *P. veris* 'Sunset Shades' (below)

P. veris (cowslip primrose) forms a rosette of deeply veined, crinkled foliage. Small clusters of tubular yellow flowers are borne at the tops of narrow stems. The plant grows about 10" tall, with an equal spread.

P. vulgaris (English primrose, common primrose) grows 6–8" tall and 8" wide. The flowers are solitary and are borne at the ends of short stems that are slightly longer than the leaves.

Problems & Pests

Slugs, strawberry root weevils, aphids, rust and leaf spot are possible problems for primroses.

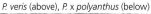

P. veris (above), *P.* x *polyanthus* (below)

Purple Coneflower

Echinacea

Also called: Coneflower, Echinacea

Height: 18"–5' **Spread:** 12–24" **Flower color:** purple, pink, yellow, orange or white, with rust orange centers **Blooms:** mid-summer to fall **Zones:** 3–8

BREEDERS HAVE GONE CUCKOO OVER CONEFLOWERS AS OF LATE, creating all manner of new colors and forms—mango, peach and raspberry, doubles, short ones, petals that are wider, shorter or tip downward. It's hard not to want some, even if you never cared for them before. Purple coneflowers are practically indestructible because they are sub-zero hardy and withstand heat, grow in dry locations and endure poor soil. They lure butterflies and birds, especially goldfinches and chickadees. They are suitable as dry flowers, excellent as cut flowers and make fine container specimens. They're good for naturalizing a large area. These plants are long lived, low maintenance and easy to grow. They're supposedly deer resistant, but someone chomped the heads off 'Art's Pride' last summer and I have my suspicions.

Planting

Seeding: Direct sow in spring

Planting out: Spring

Spacing: 12–24"

Growing

Purple coneflower grows well in **full sun** or **very light shade**. Any well-drained soil is tolerated, though an **average to rich soil** is preferred. The thick taproot makes this plant drought resistant, but it prefers to have regular water. Division can be difficult because of the stout tap-root, but small offsets are likely to develop at the base of the plant and these can be removed to propagate more plants.

Deadhead early in the season to pro-long flowering. Later you may wish to leave the flowerheads in place to self-seed and to provide winter interest. To prevent self-seeding, remove all the flowerheads as they fade. Pinch plants back or thin out the stems in early summer to encourage bushy growth that is less prone to mildew.

Tips

Use purple coneflower in meadow gardens and informal borders, either in groups or as single specimens. The dry flowerheads make an inter-esting feature in fall and winter gardens.

Because the base of the flower is rather prickly, it was was derived from echinos, *the Greek word for "hedgehog."*

E. purpurea 'White Lustre' (above), *E. purpurea* (below)

E. purpurea 'Magnus' and 'White Swan' (above)
E. purpurea 'Magnus' (below)

Recommended

E. **'Double Decker'** has deep rose-colored flowers. Out of the central cone grows a second set of petals, producing a madcap effect. It grows up to 40" tall and spreads about 24".

E. **'Fragrant Angel'** has fragrant, white flowers. It grows about 36" tall and spreads about 24".

E. **'Harvest Moon'** has golden yellow petals surrounding an orangy yellow center. It grows about 24" tall, with an equal spread.

E. **ORANGE MEADOWBRIGHT** ('Art's Pride') bears deep orange flowers with reddish orange centers. It grows about 24" tall, with an equal spread.

E. paradoxa (yellow coneflower) is a bushy, upright plant that grows 24–36" tall and spreads 18–24". It bears yellow, daisy-like flowers with brown centers.

E. purpurea is an upright plant 5' in height and up to 18" in spread, with prickly hairs all over. It bears purple flowers with orangy centers. The cultivars are generally about half the species' height. **'Kim's Knee High'** is a bushy plant much lower growing than the species. It grows 18–24" tall. **'Magnus'** bears flowers like those of the species but larger, up to 7" across. **'White Lustre'** bears white flowers with orange centers. **'White Swan'** is a compact plant with white flowers.

E. **'Twilight'** bears dark pink flowers with striking red centers. It grows about 36" tall and spreads about 24".

The Chicago Botanic Garden is field testing some exciting new hybrids with flowers in shades of deep pink, peach, sunset orange and yellow. Keep an eye open for them.

Problems & Pests

Powdery mildew is the biggest problem for purple coneflower. Also possible are leaf miners, bacterial spot and gray mold. Vine weevils may attack the roots.

E. Orange Meadowbright (above), *E. paradoxa* (center)

Echinacea is an important traditional medicine for Native Americans, and it has become a popular immunity booster in modern herbal medicine, although its claim to be a cure for the common cold has been refuted recently.

E. 'Fragrant Angel' (below)

Sage
Salvia

Also called: Perennial Salvia

Height: 12–36" **Spread:** 18–36" **Flower color:** purple, blue, pink, cream **Blooms:** late spring, summer, early fall **Zones:** 3–9

THERE ARE OVER 900 SPECIES OF *SALVIA*, SOME OF WHICH ARE annuals while others are biennials and perennials. Spiky sages make nice filler plants in borders, useful to play against the mounded forms of other plants and hide the gangly stalks of leggy shrubs. The culinary sage, *S. officinalis*, has a more rounded shape, and the prettiest of all, *S. argentea*, looks like a giant, wooly, silver cabbage. I always try to plant the latter in groups of three or more, and cut off the flower stalks before they form to keep the leaves going as long as possible. Silver sage is not grown for its flowers. Sage is usually ignored by deer. I find the pungent smell of sage somewhat unpleasant, and I guess the deer do, too.

Planting

Seeding: Cultivars do not come true to type; species can be started in early spring

Planting out: Spring

Spacing: 18–24"

Growing

Salvias prefer **full sun** but tolerate light shade. The **soil should be of average fertility and well drained.** These plants benefit from a light mulch of compost each year. They are drought tolerant once established. Division can be done in spring, but the plants are slow to re-establish and resent having their roots disturbed. They are easily propagated by tip cuttings.

Deadhead to prolong blooming. Trim plants back in spring to encourage new growth and keep plants tidy. New shoots will sprout from old, woody growth.

Tips

All *Salvia* species are attractive plants for the border. Taller species and cultivars add volume to the middle or back of the border, and the smaller specimens make an attractive edging or feature near the front. Perennial salvia can also be grown in mixed planters.

S. officinalis *has aromatic foliage that is used as a flavoring in many dishes.*

S. x *sylvestris* 'Maynight' (above)
S. officinalis 'Icterina' (below)

S. officinalis 'Icterina' (above)
S. nemerosa 'East Friesland' (center)

S. x sylvestris 'May Night' (below)

Recommended

S. nemorosa (*S.* x *superba;* perennial salvia) is a clump-forming, branching plant with gray-green leaves. It grows 18–36" tall and spreads 18–24". Spikes of blue or purple flowers are produced in summer. **'East Friesland'** is a compact plant with purple flowers. **'Lubeca'** bears long-lasting, reddish purple flowers. (Zones 3–7)

S. officinalis (common sage) is a woody, mounding plant with soft gray-green leaves. It grows 12–24" tall and spreads 18–36". Spikes of light purple flowers appear in early and mid-summer. **'Berggarten'** ('Bergarden') has silvery leaves about the size and shape of a quarter. **'Icterina'** ('Aurea') has foliage with yellow margins. **'Purpurascens'** has purple stems. The new foliage emerges purple and matures to purple-green. **'Tricolor'** has green or purple-green foliage outlined in cream. New growth emerges pinkish purple. It is the least hardy of the variegated sages. (Zones 4–7)

S*. x *sylvestris (perennial salvia) grows 30–36" tall and about 12" in spread. It is often confused with the very similar *S. nemorosa*. Cultivars have been listed under both species at different times. **'Blue Queen'** bears dark purple-blue flowers. **'May Night'** bears deep purple blue flowers. **'Rose Queen'** bears unique, rosy purple flowers, but the growth is somewhat floppier than that of the other cultivars. (Zones 3–7)

***S*. *verticillata* 'Purple Rain'** is a low, mounding plant that grows 18" tall, with an equal spread. It bears purple blooms in late summer and early fall. The colorful bracts remain long after the flowers fade. (Zones 5–8)

Problems & Pests
Scale insects, whiteflies and root rot (in wet soils) are the most likely problems.

S. nemorosa 'Lubeca' (above)
S. x *sylvestris* 'Blue Queen'

The genus name Salvia *comes from the Latin* salvus, *"save," referring to the medicinal properties of several species.*

Sea Holly

Eryngium

Height: 1–5' **Spread:** 12–24" **Flower color:** purple, blue, white
Blooms: summer, fall **Zones:** 4–8

NOT FOR THE FAINT OF HEART, SEA HOLLIES MAKE FOR A DYNAMIC
and daunting lot. From leathery, lobed leaves to spiny, decorative bracts, this
may be a plant you won't want to tangle with, especially with intimidating
nicknames such as "rattlesnake master." Nevertheless, stiff-legged sea hollies
are interesting in a late summer garden where they're at home in arid, hot
and sandy places, perfect for the beach house. They'll grow nicely alongside
purple coneflower (*Echinacea*). The flowers dry into seedheads that are
equally unusual ornaments in the winter landscape.

Planting

Seeding: Not recommended, though
direct sowing in fall will produce
spring seedlings

Planting out: Spring

Spacing: 12–24"

Growing

Grow sea hollies in **full sun**. The soil should be **average to fertile and well drained**. These plants have a long taproot and are fairly drought tolerant, but they will suffer if left more than two weeks without water. Sea hollies are very slow to re-establish after dividing. Root cuttings can be taken in late winter.

The leaves and flower bracts of these plants are edged with small spines, making deadheading a pain—literally. It is not necessary unless you are very fussy about keeping plants neat.

Tips

Mix sea hollies with other late-season bloomers in a border, or add them to naturalized gardens.

Recommended

E. alpinum (alpine sea holly) grows 2–4' tall. This species has soft and feathery-looking but spiny bracts, and its flowers are steel blue or white. Several cultivars are available in different shades of blue.

E. bourgatii forms a clump of dark green leaves with silver veins. It grows 6–18" tall and spreads about 12". It bears blue or gray-green flowers in mid- and late summer.

E. x tripartitum grows 24–36" tall. The flowers are purple and the bracts are gray, tinged with purple.

E. varifolium (Moroccan sea holly) grows 12–16" tall. It has dark green leaves with silvery veins and gray-purple flowers with blue bracts.

E. yuccifolium (rattlesnake master) is native to central and eastern

E. alpinum (above), *E. yuccifolium* (below)

North America. It grows 3–4' tall and forms a rosette of narrow, 36" long, spiny, blue-gray leaves. From mid-summer to fall, creamy green or pale blue flowers with gray-green bracts that look like groups of spiky 1" golf balls appear. Native Americans brewed a tea of the root as an antidote to rattlesnake venom, hence the intriguing common name.

Problems & Pests

Roots may rot if the plants are left in standing water for long periods of time. Slugs, snails and powdery mildew may be problems.

Sedge
Carex

Height: 12–36" **Spread:** 12–36" **Flower color:** brown **Blooms:** late spring, summer **Zones:** 6–9

ALTHOUGH MANY OF THE 1,000 SPECIES OF *CAREX* MAY LOOK LIKE grasses—the family that includes wheat, barley, rice, sugar and oats—sedges are actually from a different family of mostly marsh-dwelling plants that includes Chinese water chestnuts, bulrushes and papyrus. Regardless, many of these plants prosper with normal moisture circumstances. I grow the ones recommended here in my very dry garden. *C. buchananii* forms a gracefully sweeping rust-color tuft in full sun at my place. It also makes a perfect container specimen. Bowle's golden sedge garners attention from a distant bed—a virtual spotlight. 'Evergold' arches beautifully over the edges of a container in bright sun. The strappy leaves of the variegated, broad-leaved sedge would never be mistaken for a grass. It seems happy in full shade, and it suffered (but recovered) during drought.

Planting

Seeding: Start seed in a cold frame in fall, except leather-leaf sedge, which can be started in spring

Planting out: Spring or fall

Spacing: 12–24"

Growing

Sedges grow well in **full sun** and **partial shade**. The soil should be **fertile, moist and well drained**, though Bowles' golden and broad-leaved sedges will grow well in wet soil. Cut back old growth or thin by hand in spring to make room for the new growth. Divide overgrown clumps in early summer.

Tips

Add sedges to beds and borders in damp areas of the garden. Use them around the borders of a water feature, or directly in the water for water-loving species.

Recommended

C. buchananii (leatherleaf sedge) is a dense, tuft-forming evergreen sedge with arching, orange-brown, grass-like foliage and spikes of small, brown flowers in mid- to late summer. It grows 24–30" tall and up to 36" wide.

C. elata '**Aurea**' (Bowles' golden sedge) forms a clump of arching, grass-like, yellow leaves with green margins. It grows 24–36" tall and spreads about 18". Spikes of tiny, brown or green flowers appear in early summer. Try it with purple-leaved coral bells (*Heuchera*) or ligularia (*Ligularia*) 'The Rocket.'

C. oshimensis (*hachijoensis*) '**Evergold**' (variegated Japanese sedge) forms a low mound of evergreen, grass-like, dark green-and-yellow-striped foliage. It grows about 18" tall and spreads about 12", bearing spikes of small brown flowers in late spring and early summer.

C. oshimensis 'Evergold' (above), *C. elata* 'Aurea' (below)

C. siderosticha '**Variegata**' (variegated broad-leaved sedge) forms a clump of pale green, wide, strap-shaped leaves with creamy margins and pink-flushed bases. It grows about 12" tall and spreads about 18", producing spikes of light brown flowers in late spring.

Problems & Pests

Rare problems with aphids, fungal leaf spot, rust and smut can occur.

Sedum

Sedum

Also called: Stonecrop

Height: 2–24" **Spread:** 18" to indefinite **Flower color:** yellow, white, red, pink; plant also grown for foliage **Blooms:** summer, fall **Zones:** 3–8

SEDUMS, LIKE HENS AND CHICKS (*SEMPERVIVUM*), MAKE indispensable plants for dry, sunny gardens and can be used in very creative ways. I grow gold moss stonecrop (*S. acre*) in a shallow flat on the diving-board of my pool. The plants overflow the edges and "drip" almost to the water surface. Needless to say, I don't dive, but the board always elicits favorable comment, a lot more than I'd ever get attempting a triple summersault, one-and-a-half jackknife backflip off the board. Consequently, however, the overly assertive *S. acre* has escaped and insinuated itself all over my garden. But it's an attractive groundcover and easy enough to apprehend and put back in its place. Other species are fine in borders, containers, rock gardens, rooftop gardens, between stepping stones, or in walls, depending on their height and nature.

Planting

Seeding: Sow indoors in early spring. Seed sold is often a mix of different species; you may not get what you expected, but you may be pleasantly surprised.

Planting out: Spring

Spacing: 18"

Growing

Sedums prefer **full sun** but tolerate partial shade. The soil should be of **average fertility, very well drained and neutral to alkaline**. Divide in spring when needed. Prune back 'Autumn Joy' in May by one-half and insert pruned-off parts into soft soil; cuttings root quickly. Early-summer pruning of upright species and hybrids gives compact, bushy plants but can delay flowering.

S. spurium 'Dragon's Blood' (above)
S. kamtschaticum (below)

Tips

Low-growing sedums make wonderful groundcovers and additions to rock gardens or rock walls. They also edge beds and borders beautifully. The taller types give a lovely late-season display in a bed or border.

Low-growing sedums make excellent groundcovers under trees. Their shallow roots survive well in the competition for space and moisture.

S. acre (above), S. spectabile 'Brilliant' (below)

Recommended

S. acre (gold moss stonecrop) grows 2" high and spreads indefinitely. The small, yellow-green flowers are borne in summer. Plant this species only if you want it forever.

S. 'Autumn Joy' (*S.* 'Herbstfreude') is a popular upright hybrid. The flowers open pink or red and later fade to deep bronze over a long period in late summer and fall. The plant forms a clump 24" tall, with an equal spread.

S. kamtschaticum forms a clump with glossy green leaves. It grows about 4–8" tall and spreads 12–24". It produces bright yellow flowers in late summer.

S. spectabile (showy stonecrop) is an upright species with pink flowers borne in late summer. It forms a clump 16–24" tall and wide. **'Brilliant'** bears bright pink flowers. **'Neon'** also has deep rosy pink flowers.

S. spurium (two-row stonecrop) forms a mat about 4" tall and 24" wide. The mid-summer flowers are deep pink or white. **'Dragon's Blood'** ('Schorbuser Blut') has bronze- or purple-tinged foliage and dark pink to dark red flowers. **'Fuldaglut'** bears red or rose pink flowers above orange-red or maroon foliage. **'John Creech'** grows half the size of the species with tiny, scalloped leaves and bright pink, fall flowers. **'Royal Pink'** has dark pink flowers and bright green foliage.

S. ternatum (woodland stonecrop) forms a low, irregular mound of glossy green leaves in partial to full shade. It grows about 8" tall, spreads 12–24" and produces white flowers from mid- to late spring.

S. 'Vera Jameson' is a low, mounding plant with purple-tinged stems and pinkish purple foliage. It grows up to 12" tall and spreads 18". Clusters of dark pink flowers are borne in late summer and fall.

Problems & Pests

Slugs, snails and scale insects may cause trouble for these plants. Plants may rot in wet locations.

'Autumn Joy' brings welcome color to the late-season garden, when few flowers are in bloom.

S. spectabile 'Brilliant' (above), *S. spectabile* hybrid (below)

Shasta Daisy

Leucanthemum

Height: 10–40" **Spread:** about 24" **Flower color:** white with yellow centers
Blooms: early summer to fall **Zones:** 4–9

SHASTA DAISIES RESULTED FROM AMERICAN PLANTSMAN LUTHER
Burbank's effort to create the perfect cut flower. In 1890 he succeeded, at the
same time creating a remarkable, drought-resistant, cold-hardy, long-bloom-
ing, frilly, white winner with many landscaping virtues. Ideal for the middle
of a border and for foundation plantings, it is compatible with many grasses
for a mixed meadow or prairie composition, and is a country garden staple.
Nowadays there are many pretty cultivars. A useful tip for Shasta daisy care is
to leave the crowns uncovered in winter. Plants like coral bells, Stokes' aster
and others that maintain an above-ground presence with foliage that does
not completely disappear over winter usually resent having their heads
smothered under mulch. They will do fine without a winter hat. I lost several
perfectly nice Shasta daisies clumps before I learned this lesson.

Planting

Seeding: Start seeds indoors or
direct sow in spring with soil
temperature at about 70°–75° F

Planting out: Spring

Spacing: 24"

Growing

Shasta daisy grows well in **full sun** or **partial shade**. The soil should be **fertile, moist and well drained**. Divide every year or two, in spring, to maintain plant vigor. Fall-planted Shasta daisy may not become established in time to survive the winter. Plants can be short-lived in Zones 4 and 5.

Pinch or trim plants back in spring to encourage compact, bushy growth. Deadheading extends the bloom by several weeks. Do not cover the crowns of Shasta daisies or other plants with evergreen foliage with mulch in winter.

L. x *superbum* (all photos)

Tips

Use this perennial in the border, where it can be grown as a single plant or massed in groups. The large, showy flowerheads—some 2–5" across—can be cut for fresh arrangements.

Recommended

L. x *superbum* forms a large clump of dark, evergreen leaves and stems. It bears white, daisy-like flowers with yellow centers all summer and often until the first frost, especially if deadheaded. **'Alaska'** sends out large flowers and is hardier than the species. **'Becky'** grows about 36" tall and produces many single flowers. It is drought tolerant and one of the best choices for hot summer climates. **'Marconi'** has large, semi-double or double flowers. It should be protected from the hot afternoon sun. **'Silver Spoon'** grows to about 40" tall and bears spidery flowers with long, narrow petals. **'Snow Lady'** is a dwarf 10–14" tall with single flowers. **'Wirral Supreme'** is 36" tall and needs staking. It has large, double, white flowers with creamy yellow centers.

Problems & Pests

Occasional problems with aphids, leaf spot, leaf miners and powdery mildew are possible.

Solomon's Seal

Polygonatum

Height: 8"–5' **Spread:** 12–24" **Flower color:** white or greenish white
Blooms: spring or summer **Zones:** 4–8

SOLOMON'S SEAL SENDS UP ITS ARCHING STEMS FROM KNOBBY rhizomes that creep below the soil's surface. Each year's growth leaves a circular scar on the rhizome (called a cicactrix), which someone decided resembled the seal on King Solomon's ring, bestowing upon *Polygonatum* its best-known common name. Because of this seal-shaped scar, the plant was presumed capable of sealing wounds and broken bones, among other miracle performances. In fact, today I found a website hawking "Solomon's Seal Root Chips to increase wisdom and make wise decisions." (Wowee! Sounds like something I can use!) There are currently about 60 recognized species of *Polygonatum*. Robust and long-lived, Solomon's seal is a nice woodland accompaniment for boldly textured plants, such as hostas, mayapples, arum and Jack-in-the-pulpit, or have it spraying up through a bed of ferns, wild ginger or Siberian bugloss. Solomon's seal can even be grown in containers.

Planting

Seeding: Start seed in a cold frame in fall

Planting out: Spring or fall

Spacing: 12–24"

Growing

Solomon's seal grows well in **light shade, partial shade** or **full shade**. Soil should be **fertile, humus rich, moist and well drained**. Carefully divide plants in spring or fall. New shoots can easily be damaged in spring.

Tips

These lovely woodland plants are suitable for shaded beds and borders as well as rock gardens and woodland gardens. Their arching habit makes them attractive when grown in large clumps, especially near a water feature where they can arch across the water softening the edges of a stream or pond.

Recommended

P. biflorum forms an upright mass of arching stems. It grows 2–5' tall and spreads about 24". White flowers dangle below each leaf joint from late spring to mid-summer.

P. humile is a low-growing, spreading species that grows 8–12" tall with a spread of about 24". It bears white flowers in late spring.

P. odoratum (fragrant Solomon's seal) forms a mass of arching stems with alternating leaves forming two rows along its length. Small, white, fragrant flowers dangle from the stem at each leaf joint in late spring.

P. odoratum (above & below)

It grows up to 36" tall with a spread of about 12". **'Variegatum'** has creamy margined leaves.

Problems & Pests

Slugs and sawfly larvae may damage foliage.

Speedwell

Veronica

Also called: Veronica

Height: 6"–6' **Spread:** 12–24" **Flower color:** blue, white, pink, purple
Blooms: late spring, summer, fall **Zones:** 3–8

THERE ARE ABOUT 250 *VERONICA* SPECIES, BUT ONLY ABOUT 12
make suitable garden plants. Some are mat-forming creepers and others are
upright growers 36" high. Many make nice container plants and are consider-
ably drought tolerant. *V. peduncularis* 'Georgia Blue' is a billowy, low-growing gem, smothered in bright, true-blue blossoms early in spring with dark
green leaves that turn bronze in fall. It makes an excellent rock garden plant.
I would cover my whole garden with this if I could find enough. It's also a
great foil for daffodils, tulips and other spring-flowering bulbs. Later on in
the season, I enjoy *V.* 'Sunny Border Blue'. It forms a dense clump of spikes
that blooms for a good eight weeks from
mid-summer through fall, with
mildew-resistant foliage that turns
yellow before frost. It makes a
good cut flower. Try it
with *Coreopsis*
'Moonbeam.'

Planting

Seeding: Not recommended; seedlings do not always come true to type. Seeds germinate quickly when started indoors in early spring.

Planting out: Spring

Spacing: 18"

Growing

Speedwells prefer **full sun** but tolerate partial shade. The soil should be of **average fertility, moist and well drained**. Once established, speedwell tolerates short periods of drought. Lack of sun and excessive moisture and nitrogen may be partly to blame for the sloppy habits of some speedwells. Frequent dividing ensures strong, vigorous growth and decreases the chances of flopping. Divide in fall or spring every two or three years.

When the flowers begin to fade, remove the entire spike where it joins the plant to encourage rapid repeat blooming. For tidy plants, shear back to 6" in June.

Tips

Hungarian speedwell is a beautiful plant for edging borders. Prostrate speedwell is useful in a rock garden or at the front of a perennial border. Spike speedwell works well in masses in a bed or border.

Recommended

V. austriaca subsp. *teucrium* (Hungarian speedwell) is a clump-forming plant that grows 6–24" tall and spreads 12–24". It bears spikes of bright blue flowers from late spring to midsummer. **'Crater Lake Blue'** grows 12–18" tall and bears deep blue flowers.

V. 'Sunny Border Blue' (above & below)

V. spicata 'Red Fox' (above)
V. spicata subsp. incana (below)

V. **'Goodness It Grows'** forms a low, spreading mound. It grows about 12" tall with a spread of about 18". It bears spikes of dark blue to purple-blue flowers all summer.

V. peduncularis is a low-growing, mat-forming plant with glossy, purple-tinged leaves that turn bronzy in fall. It grows about 4' tall and spreads at least 24". It bears white-centered, blue flowers from spring to mid-summer. **'Georgia Blue'** produces lots of striking, sapphire blue flowers. **'Waterperry Blue'** has light blue to lilac blue flowers.

V. prostrata (prostrate speedwell) is a low-growing, spreading plant 6" tall and 16" wide. Its flowers may be blue or occasionally pink. Many cultivars are available.

V. spicata (spike speedwell) is a low, mounding plant with stems that flop over when they get too tall. It grows 12–24" tall, spreads 18" and produces spikes of blue flowers in summer. Many cultivars of different colors are available. **'Blue Bouquet'** is a low-growing cultivar, to 12", with spikes of bright blue flowers. **'Blue Charm'** is a bushy cultivar growing 12–18" tall. It bears spikes of light purple-blue flowers. **'Icicle'** ('White Icicle') produces spikes of white flowers. **Subsp.** *incana* (silver speedwell) has silvery blue-green foliage and bright blue flowers. **'Red Fox'** has dark red-pink flowers.

V. **'Sunny Border Blue'** has attractive 18–24" blue-violet spikes and dark green foliage that turns yellow in fall. Deadhead to encourage repeat blooming.

Problems & Pests

Problems with scale insects are possible, as are fungal problems such as downy and powdery mildew, rust, leaf smut and root rot.

One speedwell species (V. officinalis) was substituted for tea in Europe until the 19th century. Brewed from the dried flowering plant, it is still used today by herbalists as a cough remedy or as a skin lotion to speed wound healing and relieve itching. In fact, the speed with which the plant affects cures is supposedly the reason for its common name "speedwell."

V. spicata subsp. *incana* (above), *V. spicata* cultivar (below)

Stokes' Aster

Stokesia

Height: 12–24"   **Spread:** 12–18"   **Flower color:** purple, blue, white, pink, pale yellow   **Blooms:** mid-summer to early fall   **Zones:** 4–9

A PERSONAL FAVORITE OF MINE FOR ITS EASYGOING NATURE, long-lasting, late-season bloom, pest resistance and evergreen foliage, Stokes' asters come in a variety of colors. They self-sow politely and are highly adaptable to soil and weather conditions. Most grow no taller than 24" and belong toward the front of a border. 'Omega Skyrocket,' though, is double that height. 'Colorwheel' blooms earlier than most, starting in mid-May. Its flowers open white, and over several days turn pink, then lavender, eventually deepening to darker purple. During these stages new white blooms open on the same stalk, showing as many as five different colors on the same plant. Each blossom retains a white center. Novel, pretty and entertaining, too! Stokes' asters make excellent cut flowers, staying open for a week or more. Butterflies love them. Deer seem disinterested. They make good container plants. What more could you ask?

Planting

Seeding: Start seeds indoors or out in fall with soil temperature at 70–75° F

Planting out: Spring

Spacing: 12"

Growing

Stokes' aster grows best in **full sun**. The soil should be **average to fertile, light, moist and well drained**. This plant dislikes waterlogged and poorly drained soils, particularly in winter when root rot can quickly develop. Those who garden in clay might consider planting this species in raised beds filled with carefully amended soil. Provide a good winter mulch to protect the roots from the cycles of freezing and thawing, particularly in Zones 4 and 5. Divide in spring.

Deadheading extends the bloom, which can then last up to 12 weeks.

Tips

Stokes' aster can be grouped in borders and adds welcome blue to the garden late in the season when yellows, oranges and golds seem to dominate.

Recommended

S. laevis forms a basal rosette of bright green, narrow foliage. The midvein of each leaf is a distinctive pale green. The plant bears purple, blue or sometimes white or pink flowers from mid-summer to early fall. **'Alba'** has white flowers. **'Blue Danube'** bears large purple-blue flowers. **'Colorwheel'** produces flowers that open white and mature to deep purple, showing every color in between. Flowers in shades of pink, purple and white appear, sometimes all at once on the plant, creating a virtual tapestry. **'Klaus Jelitto'** bears large, light blue flowers. **'Mary Gregory'** is a compact cultivar with light yellow flowers. **'Omega Skyrocket'** is a strong-growing, 3–4' tall plant that

S. laevis (all photos)

bears large purple flowers. **'Peachie's Pick'** is a dense-growing, compact plant that bears blue flowers. **'Purple Parasols'** has flowers that open blue and mature to purple. **'Rosea'** produces pink flowers.

Problems & Pests

Problems are rare, though leaf spot and caterpillars can cause problems. Avoid very wet and heavy soils to prevent root rot.

Sunflower

Helianthus

Height: 3–8' **Spread:** 36" **Flower color:** yellow and brown **Blooms:** late summer, fall **Zones:** 4–8

THE SMILEY FACE SUNFLOWERS ARE ALL NATIVE TO THE AMERICAS. Domesticated during prehistoric times, Native Americans recognized their food and medicinal value, and venerated them in their religious ceremonies. All plant parts were used: to cure rattlesnake bites, soothe chest pains and cure wounds; to be smoked like tobacco, brewed into a coffee-like drink, made into cloth and dyes; and, of course, to eat. After the discovery of America, Spaniards introduced them to Europe, where they were welcomed with particular enthusiasm in Russia, and developed into a major crop. Farmers fed livestock the seeds, miners drank sunflower whiskey and Jerusalem artichokes (*H. tuberosum*) fed both animals and humans. A nuisance weed in corn and soybean fields, sunflower rivals its competitors in the culinary oil market. Not simply a major economic crop, sunflowers are unbeatable for late-summer garden entertainment, when bees and butterflies hover and goldfinches sit atop the stalks and feast on the seeds.

Planting

Seeding: Start in spring; seedlings may not come true to type

Planting out: Spring

Spacing: 24–36"

Growing

Sunflowers grow best in **full sun**. They flower less and become lanky in shaded conditions. Ideally, the soil should be of **average fertility, neutral to alkaline, moist and well drained**. *H.* x *multiflorus* prefers a constantly moist soil, while *H. salicifolius* is drought tolerant.

Plants should be cut back hard after flowering. They can also be cut back in early summer to produce shorter but later-flowering plants. Divide every three or so years in spring or fall.

Tips

These impressive perennials deserve a spot at the back of the border or in the center of an island bed. Sunflowers are tall plants that can provide a quick privacy screen in an exposed garden. Water-loving *H.* x *multiflorus* can also be planted near a pond or other water feature. *H. salicifolius* can be used in dry and underwatered areas of the garden.

Recommended

H. '**Capenoch Star**' forms a large clump of stems with narrow leaves. It grows 4–5' tall and spreads 3–4'. It bears light yellow flowers from mid-summer to early fall.

H. '**Lemon Queen**' has single, bright but soft lemon yellow blooms on 6–8' tall, sturdy, upright stems. It blooms from mid-summer to frost.

H. '**Loddon Gold**' bears golden yellow, double flowers.

H. x *multiflorus* (perennial sunflower) forms a large, upright clump 3–6' tall and about 36" in spread.

H. salicifolius (above), *H.* 'Loddon Gold' (below)

Many daisy-like flowers with yellow petals and brown centers are borne in late summer and fall. '**Soleil d'Or**' bears yellow, double flowers.

H. salicifolius (willow-leaved sunflower) is a large, clump-forming plant with narrow leaves. It grows up to 8' tall, spreads 36" and bears daisy-like, yellow flowers with brown centers in fall.

Problems & Pests

Rare problems are possible with powdery mildew, downy mildew, fungal leaf spot and leaf-chewing insects like caterpillars, beetles and weevils.

Toad Lily

Tricyrtis

Height: 24–36" **Spread:** 12–24" **Flower color:** white, pink or purple with red or purple spots. **Blooms:** late summer and fall **Zones:** 4–9

SOME ATTRIBUTE THIS FASCINATING SPECIES' COMMON NAME TO a story alleging that the Philippine Tasaday Indians coat their arms and hands with *Tricyrtis* juice before setting out to catch frogs, the tribe's only source of protein. Supposedly, the scent is an attractant and the sticky juice makes them easier to grip. The spotted, orchid-like flowers are possibly the true justification for the name. Regardless, the understated beauty and intricacy of the dainty blossoms dictate that toad lilies be planted en masse where they can be admired close-up. Blooming on gently arching stems, toad lilies are an acknowledged treat in late summer and early fall when few flowers bother. The showiest one in my garden is T. *hirta* 'Moonlight'. Its neon chartreuse leaves really light up the woodland.

Planting

Seeding: Not recommended

Planting out: Spring

Spacing: 12–24"

Growing

Toad lilies grow well in **partial shade, light shade** or **full shade**. The soil should be **fertile, humus rich, moist and well drained**.

Tips

These plants are well suited to plantings in woodland gardens and shaded borders. If you have a shaded rock garden, patio or pond, these plants make good additions to locations where you can get up close to take a good look at the orchid-like, spotted flowers.

Recommended

T. 'Empress' forms a large, upright clump growing about 30" tall and spreading 18–24". It bears large, white flowers with dark, reddish purple spots.

T. formosana (formosa toad lily) is an upright spreading plant. It grows 24–36" tall and spreads 18–24". The early fall flowers are white, sometimes tinged pink or purple, with red or purple spots. 'Amethystina' bears pale purple flowers with white throats and red spots. 'Samurai' is a compact plant with yellow-margined leaves and light purple flowers with darker purple spots.

T. hirta (Japanese toad lily) forms a clump of light green leaves. In late summer and fall, it bears purple-spotted, white flowers. It grows 24–36" tall and spreads 12–24". 'Miyazaki' has white flowers that are spotted with light purple. The leaves have lighter margins. 'Moonlight', which really brightens up the shade, is a golden-leaved

T. hirta

sport of 'Variegata'. It grows 20" tall with white flowers flecked occasionally with lavender. 'Variegata' has narrow, creamy edges on the medium green foliage of 15" tall plants.

T. 'Moonlight Treasure' is a compact form with olive green, faintly spotted, thick, leathery leaves, very densely set along the 12" stem. The buttery yellow flowers are large in relation to the plant, and face upward at the leaf axils, growing in clusters of two to five.

T. 'Sinonome' forms a dense clump of stems and leaves. It grows up to 36" and spreads 18–24". Its flowers are purple-and-white spotted.

Problems & Pests

Slugs can cause damage in spring as the new growth emerges.

Trillium

Trillium

Also called: Wake Robin

Height: 16–20" **Spread:** 12" or more **Flower color:** white, yellow, pink, red, purple **Blooms:** spring **Zones:** 4–7

DELIGHTFUL RED, WHITE AND PAINTED TRILLIUMS ARE ALL FAIRLY common in the wild in Pennsylvania. Red trillium can be a startling maroon red, or appear in various paler shades of white, pink or mauve. Great white trillium is a large white jewel that turns pink as it ages. Strikingly beautiful, painted trillium is a worthwhile challenge to grow in cultivation. Yellow trillium has a pleasant citrus scent and lovely mottled leaves. Toadshade trillium has upright, maroon flowers and spotted foliage. Spent flowers are followed by a single fruit, which ants cart away to their underground lairs, eating its pulp and then, in essence, replanting the seeds. These take many years to become a flowering-sized bulb.

Planting

Seeding: Not recommended; ripe seeds may be started in a shaded cold frame in late summer.

Planting out: Fall or spring

Spacing: 12"

Growing

Locate trilliums in **full** or **partial shade**. The soil should be **humus rich, moist, well drained and neutral to acidic**. Rhizomes should be planted about 4" deep. Add organic matter, such as compost or aged manure, to the soil when planting, and add a mulch of shredded leaves to encourage rapid growth. Division is not necessary. Apply compost every year.

Tips

These plants are ideal for natural woodland gardens and for plantings under spring-flowering trees and shrubs.

Trilliums are best left alone once planted. New transplants may take a year or two to adjust and start flowering. Plentiful moisture in summer prevents the plants from going dormant after flowering and helps them establish quickly. They send up side shoots that increase the size of the clump and the number of flowers the following spring.

Recommended

T. erectum (purple trillium, red trillium) has wine red flowers. It grows up to 20" tall and spreads up to 12". When the flower is spent, around June, it produces a red berry.

T. grandiflorum (great white trillium, snow trillium) has large, white flowers that turn pink as they mature. It grows 16" tall and spreads 12" or more. When the flower dies back, a black fruit follows. **'Flore Pleno'** has double flowers but is slower growing.

T. luteum (yellow whippoorwill, golden goblet trillium) forms a clump of dark green leaves spotted with lighter green. It bears fragrant, bright yellow flowers in spring. Plants grow about 16" tall and spread about 12".

T. sessile (toadshade trillium) forms a clump of dark green leaves mottled with sliver, bronze, maroon and lighter green. It bears dark purple-red flowers in late spring. Plants grow about 12" tall and spread about 8".

T. undulatum (painted trillium) has pure white, upward-facing flowers with a prominent triangle of red-purple in the throat that bleeds up along the veins of the petals to the tips. The fruit is a distinctive,

T. erectum (above), T. undulatum (below)

smooth, oval berry with a pointed tip. This species is difficult to grow and, therefore, harder to find in nurseries compared with the others. It is endangered in many states.

Problems & Pests

Trilliums have few pest problems, but the young foliage may be attacked by slugs.

Tulip

Tulipa

Height: 6–30" **Spread:** 2–8" **Flower color:** all shades except blue **Blooms:** spring **Zones:** 3–8; sometimes treated as an annual

IF YOU ARE ANNOYED WITH TULIPS THAT DISAPPEAR AFTER A YEAR or two, consider planting "botanical" tulips, which include species tulips and their relatives: the early, large-flowering, brightly colored Fosteriana (Emperor) tulips, the low-growing, showy Kaufmanniana (water lily) tulips and Greigii tulips, known for their eye-catching, purple-striped and mottled foliage. Darwin Hybrids also perennialize well. Many of the species tulips like *T. tarda* and *T. sylvestris* spread by stolons. Others spread by seeding around. Long-lasting, later blooming *T. batalinii* hybrids, ranging from yellows and apricots to reds, are great performers. Some flower in clusters, like the yellow and white *T. tarda,* and *T. sylvestris,* blooming successively throughout spring. Some species tulips are fragrant, like *T. sylvestris,* and the orange *T.* 'Little Princess' and purple/red *T.* 'Little Beauty,' but only your cat will be low enough to notice.

Planting

Seeding: Not recommended

Planting out: Fall

Spacing: 2–8"

Growing

Tulips grow best in **full sun**. The flowers tend to bend towards the light in light or partial shade situations. Soil should be **fertile and well drained**. Plant bulbs in fall, 4–6" deep. Bulbs that have been cold treated may be available for purchase in the spring and can be planted at that time. Although many tulips will come back and bloom each spring, many hybrids perform best if planted new each year. The species and older cultivars are the best choices for naturalizing.

Tips

Tulips provide the best display when mass planted or planted in groups in flowerbeds and borders. They can also be grown in containers and can be forced to bloom early in pots indoors. Some tulips can be naturalized in cottage, meadow and wildflower gardens.

During the "tulipomania" of the 1630s, tulip bulbs were worth many times their weight in gold, and many tulip speculators lost massive fortunes when the mania ended.

T. batalinii (above), *T.* 'Purissima' (center)

Recommended

T. batalinii has gray-green leaves with wavy, reddish edges. Small, bowl-shaped flowers in various shades of yellow, apricot, and red emerge in late April to early May. They grow 8–12" tall.

T. 'Blushing Apeldorn' is a Darwin hybrid tulip. It has large, cup-shaped, lemon yellow flowers with persimmon orange edges and interior feathering. It grows 22" tall and blooms from mid-April to May.

T. 'Heart's Delight' is a Kaufmanniana tulip. It is often called the waterlily tulip because on a sunny day the petals open flat like a star, giving the appearance of a waterlily. It grows 10" high. The early April flowers are rich carmine red on the outside of the petals and white aging to pale rose on the interior, with a yellow center.

T. 'Little Beauty' has pinkish red, star-shaped, early May flowers with pale pink–lined, dark blue centers. It grows 4–6" tall.

T. 'Little Princess' has dark orange, star-shaped, early May flowers with yellow-lined, dark blue centers. It grows 4–6" tall.

T. 'Purissima' ('White Emperor') is a Fosteriana tulip with pure white, fragrant flowers with immense petals. This cultivar is fantastic planted in large drifts or used in bridal bouquets. It is 18" tall and blooms in April or May.

T. sylvestris has light green leaves and grows in sun or shade. It bears fragrant, star-shaped, yellow- or

cream-colored flowers in April. It grows about 16" tall.

T. tarda is very easy to grow in full sun. It has narrow, glossy green leaves and star-shaped white, yellow-centered flowers, sometimes five to a stem. They flower abundantly for a long time from April through May. They grow only 4–6" in height.

***T.* 'Toronto'** is a multi-flowering (several blooms per stem) Greigii tulip with purple and green-striped leaves. The bowl-shaped March or April flowers are pinkish red with orange-red insides and yellow centers. It grows 8–12" tall.

Problems & Pests

Bulb rot can occur in poorly drained soil and slugs, aphids and nematodes may attack plants. Bulbs, leaves and flowers are attacked by mice, voles, deer, rabbits and squirrels.

T. 'Heart's Delight'

Virginia Bluebells

Mertensia

Also called: Virginia Cowslip
Height: 12–24" **Spread:** 10–18" **Flower color:** blue, purple-blue
Blooms: mid- and late spring **Zones:** 3–7

NAMED IN HONOR OF EARLY GERMAN BOTANIST, FRANZ KARL
Mertens, *Mertensia* is a highly regarded native wildflower that people love to
claim as their own. Around here, Virginia bluebells are referred to as
"Brandywine" bluebells. Perhaps in your neck of the woods they have another
nickname. These members of the borage family emerge in late winter just as the
deciduous trees are beginning to leaf out, and bloom soon after. The bright
pink buds break into sky blue flowers, perfect ballerina tutus for woodland
fairy dancers. Some white or pink forms are available, but they are hard to
come by and not as vigorous. As the seed ripens, the plants go dormant. Slow
to establish, they can form large and impressive colonies once settled in. They
disappear by summer. Fill the void with hostas, ferns or any late-emerging
perennial or spreading annual.

Planting

Seeding: Start in a cold frame in fall
Planting out: Spring or fall
Spacing: 12–18"

Growing

Virginia bluebells grow best in **light shade.** Soil should be of **average fertility, humus rich, moist and well drained.** Plants will self-seed in good growing conditions. Plants go dormant in mid- to late summer and can be divided at that time or in spring, just as new growth begins.

Tips

Include Virginia bluebells in a shaded border or moist woodland garden.

Recommended

M. virginica forms an upright clump with blue-green leaves. It grows 12–24" tall and spreads 10–18". It bears clusters of blue or purple-blue flowers that open from pink buds in mid- and late spring.

Problems & Pests

Slugs may feed on young growth. Powdery mildew and rust can be a problem.

Don't forget where you planted these woodland perennials and dig them up by accident. They go dormant over the summer and don't reappear until the following spring.

M. virginica (all photos)

Wild Ginger

Asarum

Height: 3–6" **Spread:** 12" or more **Flower color:** burgundy or green, inconspicuous; plant grown for foliage **Blooms:** late spring to early summer **Zones:** 4–8

WHETHER YOU'RE USING IT AS A GROUNDCOVER OR A SPECIMEN plant, wild ginger will perform admirably, even in dense shade and clay soil. The silky textured, smoky green of *A. canadense* and the glossy, silver-veined foliage of most other gingers blends nicely with woodland shrubs and other shade dwellers. Peek under the foliage to admire the strange little flowers. They are great conversation starters, even when you're only talking to yourself. Don't limit yourself to the few listed here. Plant hunter and nurseryman Barry Yinger is responsible for finding and selling many beautiful *Asarum*, some hardy for us, some not, at his Asiatica Nursery in Lewisberry, PA.

Planting

Seeding: Start seed in a cold frame in late summer or fall; some species self-seed

Planting out: Spring or fall

Spacing: 12"

Growing

Wild gingers need **full** or **partial shade**. The soil should be **moist and humus rich**. All *Asarum* species prefer acidic soils, but *A. canadense* will tolerate alkaline conditions. Wild gingers tolerate dry conditions for a while in good shade, but prolonged drought will eventually cause the plants to wilt and die back. Division is unlikely to be necessary, except for propagation.

Tips

Use wild gingers in a shady rock garden, border or woodland garden. These plants spread to cover ground easily, but they are also relatively easy to remove from places they aren't welcome.

A. canadense (above), A. splendens (below)

You can easily propagate or share wild gingers. The thick, fleshy rhizomes grow along the soil or just under it, sprouting pairs of leaves. Cuttings can be made by removing sections of rhizome with leaves growing from them and planting each section separately. When taking cuttings, be careful not to damage the tiny, thread-like roots that grow from the stem below the leaf attachment.

Recommended

A. canadense (Canada wild ginger) is native to the Midwest and eastern North America. The heart-shaped leaves are slightly hairy. Curious little urn-shaped flowers with long mouse tails are hidden beneath the leaves in early summer.

A. europaeum (European wild ginger) is a European species with very glossy, handsome, evergreen leaves, often distinctively silver-veined. It forms an expanding clump.

A. shuttleworthii is a low, spreading, evergreen perennial with glossy, dark green, silver-marbled, heart-shaped leaves. It grows about 3" tall and spreads 12". Insignificant purple flowers are produced in early summer.

A. splendens (Chinese ginger) forms a dense spreading mass of large, 6" wide, dark green, silver-mottled evergreen leaves. It grows about 6" tall and spreads 24" or more.

Wood Fern

Dryopteris

Height: 1–4' **Spread:** 1–4' **Flower color:** no flowers **Blooms:** grown for foliage **Zones:** 3–8

IF YOU HAVE A SHADY EXPANSE TO FILL IN YOUR GARDEN, YOU MAY want to opt for creating a fern garden as an easy and attractive solution. You can select from the hundreds of species of wood ferns, which have contributed more garden-worthy ferns than any other genus. And, naturally, *Dryopteris* and its many cultivars can accompany bulbs, groundcovers, shrubs and other perennials if you choose to diversify. Not only do ferns add a lush look, but they are virtually pest free. Although I can't vouch for the critters in your neighborhood, I have never seen a rabbit, deer or insect pay the slightest attention to a fern. That alone is a practical endorsement.

Planting

Seeding: Not recommended

Planting out: Spring or fall

Spacing: 12–36"

Growing

Wood ferns grow well in **light** or **partial shade**. Soil should **fertile, humus rich** and **moist to wet,** though some species are fairly drought tolerant. Plants can be divided in spring or fall to propagate them or control their spread.

Tips

These large and impressive ferns can be used in moist, shaded areas of the garden and are at home in woodland and waterside plantings. These ferns make an ideal addition to a shaded area of the garden that floods periodically or takes a long time to dry out after it rains.

D. marginalis (above), *D. filix-mas* (below)

Recommended

D. affinis 'Cristata' (golden-scale male fern) forms a large clump of arching evergreen fronds. The leaflets on the fronds are divided at the tips, giving the plant a fluffy appearance. It grows about 36" tall, with an equal spread.

D. x australis (Dixie wood fern) forms an upright clump of semi-evergreen fronds. It grows up to 4' tall and spreads about 24".

Wood ferns are some of the easiest ferns to grow.

D. cristata (crested wood fern) is a shade-loving, wetland fern. It forms a small clump of very upright, narrow, airy fronds. Fertile fronds are deciduous and sterile fronds are evergreen. It grows 12–36" tall and spreads about 12".

D. erythrosora (autumn fern) forms tufts of shiny, dark green, deciduous fronds that contrast with the coppery-colored young fronds. It grows about 24" tall and spreads about 18".

D. filix-mas (male fern) is similar in appearance to *D. affinis*, but the fronds are lighter in color, less leathery and not evergreen. This large fern, with fronds up to 4' long, is one of the best-known wood ferns.

D. goldiana (giant wood fern), one of the largest wood ferns, spreads

D. erythrosora (above & below))

slowly, producing tufts of deciduous fronds along the rhizome. It grows about 4' tall and spreads 2–4'.

D. marginalis (leatherwood fern, marginal wood fern) forms a large clump of evergreen fronds. It grows 18–36" tall and spreads about 24".

Problems & Pests
Occasional problems with leaf gall, fungal leaf spot and rust may arise.

Dryopteris have very variable fronds. They often develop further leaflet divisions, giving the fronds a twisted or crested appearance and adding interest to your shade garden.

D. marginalis (above), *D. erythrosora* (below)

Yarrow

Achillea

Height: 4"–4' **Spread:** 12–36" **Flower color:** white, yellow, red, orange, pink, purple **Blooms:** early summer to fall **Zones:** 3–9

YARROWS ARE TROUBLE-FREE, DEER-RESISTANT, COLORFUL PLANTS that attract butterflies and sundry pollinators, appropriate for dry, sunny areas. They thrive on neglect because overwatering or fertilizing makes them floppy rather than floriferous. *Achillea* blossoms all summer, and with occasional deadheading, continues blooming well into autumn. Trimming down to the basal leaves will also induce reflowering. Many myths and superstitions follow this plant: if a woman places yarrow under her pillow and recites a particular poem, her true love's name will be revealed to her; yarrow tea taken within a church is a cure for demonic possession; by washing one's eyes with yarrow, psychic powers or second-sight are believed to be enhanced. Personally, I advise you stick to growing these merely as garden ornamentals.

Planting

Seeding: Direct sow in spring. Don't cover the seeds; they need light to germinate.

Planting out: Spring

Spacing: 12–24"

Growing

Yarrows grow best in **full sun**. The soil should be of **average fertility, sandy and well drained**. These plants tolerate drought and poor soil. They will also tolerate, but not thrive in, a heavy, wet soil or in very humid conditions. Excessively rich soil or too much nitrogen results in weak, floppy growth. Good drainage is key to getting the best growth from these plants. Divide every two or three years, in spring.

The species *A. millefolium* and *A. filipendulina* often need staking. Many of the cultivars are less floppy.

Yarrows will flower longer and more profusely if they are deadheaded. Once the flowerheads begin to fade, cut them back to the lateral buds. Basal foliage should be left in place over the winter and tidied up in spring.

Tips

Yarrows are informal plants. Cottage gardens, wildflower gardens and mixed borders are perfect places for them. They thrive in hot, dry locations where nothing else will grow.

Use yarrows in dried or fresh arrangements. It is nice to retain many of the leaves on the stems, as they have a pleasant, spicy scent. For drying, pick flowerheads just before

A. 'Moonshine' (above), A. millefolium cultivar (below)

A. millefolium 'Summer Pastels' (above)
A. 'Coronation Gold' (below)

the flowers are fully opened. For fresh bouquets, pick once the pollen is visible, or the plants will die very quickly once cut.

Recommended

A. **'Anthea'** bears bright yellow flowers that fade to creamy yellow. It will flower all summer if kept deadheaded. The foliage is silvery gray. Plants grow 12–24" tall and spread 12–18".

A. **'Coronation Gold'** has bright golden yellow flowers and fern-like foliage. It grows about 36" tall, with a 12" spread. This cultivar is quite heat tolerant.

A. **'Fire King'** has bright red flowers and fern-like foliage. It grows about 24" tall and spreads about 18".

A. x *lewisii* **'King Edward'** is a low-growing hybrid that reaches 4–8" in height and about 12" in spread. It has woolly, gray leaves and bears clusters of yellow flowers in early summer.

*A. **millefolium*** (common yarrow) grows 12–36" tall, with an equal spread, and has white flowers. The foliage is soft and very finely divided. Because it is quite aggressive, the species is almost never grown in favor of the many cultivars. **'Cerise Queen'** has pinkish red flowers. **'Summer Pastels'** has flowers of many colors, including white, yellow, pink and purple. This cultivar tolerates heat and drought very well and has fade-resistant flowers. **'Terra Cotta'** bears flowers that open orange-pink and mature to rusty orange.

A. **'Moonshine'** bears bright yellow flowers all summer. The foliage is silvery gray. The plant grows 18–24" tall and spreads 12–24".

A. ptarmica (sneezewort) grows 12–24" tall, with an equal spread. It bears clusters of white flowers all summer. **'Stephanie Cohen,'** named after one of our local horticultural heroines, bears pale pink flowers in large flowerheads. **'The Pearl'** bears clusters of white, button-like, double flowers. This cultivar can become invasive. It can be started from seed, but not all the seedlings will come true to type.

A. millefolium 'Cerise Queen' (above)
A. ptarmica 'The Pearl' (below)

Problems & Pests

Rare problems with powdery mildew and stem rot are possible.

The ancient Druids used yarrow to divine seasonal weather, and the ancient Chinese used the stems to foretell the future.

Height Legend: Low: < 12"-•-Medium: 12–24"-•-Tall: > 24"

SPECIES by Common Name	White	Pink	Red	Orange	Yellow	Blue	Purple	Foliage	Spring	Summer	Fall	Low	Medium	Tall
Ajuga	•	•				•	•	•	•	•		•		
Allium	•	•	•			•	•			•			•	•
Anemone	•	•				•	•		•	•	•	•	•	•
Artemisia	•				•			•		•	•	•	•	•
Arum	•							•		•		•		
Aster	•	•	•			•				•	•	•	•	•
Baptisia	•					•	•		•	•				•
Black-eyed Susan				•	•					•	•	•	•	•
Bleeding Heart	•	•	•					•	•	•		•	•	•
Blue Star Flower						•			•	•				•
Boltonia	•	•					•			•	•			•
Brunnera						•	•	•	•				•	
Bugbane	•	•								•	•			•
Campanula	•					•	•			•	•	•	•	•
Catmint	•	•				•	•		•	•			•	•
Chrysanthemum		•	•	•	•		•			•	•		•	•
Columbine	•	•	•		•	•	•		•	•		•	•	•
Coral Bells	•	•	•				•	•	•	•		•	•	•
Coreopsis		•		•	•					•		•	•	•
Corydalis	•				•	•	•	•	•	•		•	•	
Culver's Root	•	•				•	•			•	•			•
Cyclamen	•	•	•					•	•			•		
Daffodil	•	•		•	•				•			•	•	
Daylily		•	•	•	•		•			•			•	•
Dead Nettle	•	•			•	•	•		•	•		•	•	
Dwarf Plumbago						•				•		•	•	
Epimedium	•	•	•		•		•	•	•			•	•	
Euphorbia				•	•			•	•	•			•	
False Solomon's Seal	•									•				•
Flowering Fern								•	•	•				•
Foamflower	•	•						•	•	•		•		

| LIGHT | | | | SOIL CONDITIONS | | | | | | | | SPECIES by Common Name |
Sun	Part Shade	Light Shade	Shade	Moist	Well Drained	Dry	Fertile	Average	Poor	USDA Zones	Page Number	
	•	•			•			•		3–8	70	Ajuga
•				•	•		•	•		3–9	74	Allium
	•	•			•			•	•	5–8	76	Anemone
•					•	•		•	•	3–8	80	Artemisia
•	•			•			•			5–9	84	Arum
•				•	•		•			3–8	86	Aster
•					•	•		•	•	3–9	90	Baptisia
•	•				•			•		3–9	92	Black-eyed Susan
		•		•	•		•	•		3–9	94	Bleeding Heart
•	•	•		•	•		•	•		3–9	98	Blue Star Flower
•				•	•		•			4–9	100	Boltonia
		•		•	•			•		3–8	102	Brunnera
	•	•		•			•			3–8	104	Bugbane
•	•	•			•		•	•		3–7	106	Campanula
•	•				•			•		3–8	108	Catmint
•				•	•		•			5–9	110	Chrysanthemum
	•			•	•		•			2–9	114	Columbine
•	•			•	•		•			3–9	118	Coral Bells
•					•	•		•		3–9	124	Coreopsis
	•	•			•			•	•	5–7	128	Corydalis
•	•			•				•	•	3–8	130	Culver's Root
	•	•			•			•		5–9	132	Cyclamen
•	•			•	•		•	•		3–9	134	Daffodil
•	•	•	•	•	•		•			2–9	140	Daylily
	•	•		•	•			•		3–8	144	Dead Nettle
•	•				•			•	•	5–9	148	Dwarf Plumbago
	•	•	•	•				•	•	4–8	150	Epimedium
•		•		•	•			•		4–9	154	Euphorbia
		•	•	•	•			•		3–9	158	False Solomon's Seal
•		•		•			•			2–8	160	Flowering Fern
	•	•	•	•				•		3–8	162	Foamflower

Height Legend: Low: < 12"-•-Medium: 12–24"-•-Tall: > 24"

SPECIES by Common Name	White	Pink	Red	Orange	Yellow	Blue	Purple	Foliage	Spring	Summer	Fall	Low	Medium	Tall
Gas Plant	•	•					•			•			•	•
Globe Thistle						•	•	•		•				•
Goat's Beard	•							•		•		•	•	•
Green and Gold					•				•	•		•		
Hardy Geranium	•	•	•			•	•	•	•	•	•	•	•	•
Hardy Orchids	•	•							•	•	•		•	
Heliopsis				•	•					•	•			•
Hellebore	•	•			•	•	•		•				•	
Hens and Chicks	•		•		•		•	•		•		•		
Heucherella	•	•						•	•	•		•	•	
Hibiscus	•	•	•		•		•			•	•			•
Hollyfern								•					•	
Hosta	•						•	•		•	•	•	•	•
Indian Pink			•		•				•	•			•	
Iris	•	•	•	•	•	•	•		•	•	•	•	•	•
Jack-in-the-Pulpit						•	•		•	•			•	
Jacob's Ladder	•					•	•		•	•		•	•	•
Joe-Pye Weed	•						•			•	•			•
Lady's Mantle					•			•	•	•		•	•	
Lamb's Ears		•				•	•			•		•	•	
Ligularia				•	•			•		•	•			•
Lily	•	•	•	•	•		•			•				•
Lily-of-the-Valley	•	•							•			•		
Lilyturf						•	•	•		•	•	•	•	
Lungwort	•	•	•			•	•	•	•				•	•
Lupine	•	•			•	•	•		•	•				•
Mayapple	•		•				•	•	•	•		•	•	•
Maidenhair Fern								•					•	
Meadow Rue	•	•			•		•	•		•	•			•
Meadowsweet	•	•	•					•	•	•	•			•

Sun	Part Shade	Light Shade	Shade	Moist	Well Drained	Dry	Fertile	Average	Poor	USDA Zones	Page Number	SPECIES by Common Name
•	•				•			•		3–8	166	Gas Plant
•					•	•		•	•	3–8	168	Globe Thistle
	•		•	•			•			3–7	170	Goat's Beard
•	•			•	•			•		5–8	174	Green and Gold
•	•	•			•			•		3–8	176	Hardy Geranium
	•			•	•		•			5–8	180	Hardy Orchids
•				•	•	•	•	•	•	2–9	182	Heliopsis
		•		•	•		•			4–9	184	Hellebore
•	•				•	•		•	•	3–8	188	Hens and Chicks
•	•	•	•	•	•		•			3–8	190	Heucherella
•	•				•			•		5–8	192	Hibiscus
	•	•	•	•	•		•			3–8	196	Hollyfern
	•	•	•	•	•		•			3–8	198	Hosta
	•	•		•	•		•			6–9	202	Indian Pink
•				•	•	•	•	•		3–10	204	Iris
	•	•		•	•		•			4–9	208	Jack-in-the-Pulpit
	•	•		•	•		•			3–7	212	Jacob's Ladder
•				•	•		•			3–9	214	Joe-Pye Weed
	•	•		•	•		•			3–7	216	Lady's Mantle
•					•			•	•	3–8	220	Lamb's Ears
	•	•		•				•		4–8	222	Ligularia
•				•	•		•			4–8	224	Lily
•	•	•	•	•	•	•	•	•	•	2–7	228	Lily-of-the-Valley
•	•	•	•	•	•			•		6–9	230	Lilyturf
	•	•	•	•	•		•			3–8	232	Lungwort
•	•				•	•	•	•		3–8	236	Lupine
	•		•	•			•			5–8	238	Mayapple
	•	•	•	•	•			•		2–8	242	Maidenhair Fern
	•	•		•	•			•		3–8	244	Meadow Rue
	•	•		•		•	•			3–8	248	Meadowsweet

Height Legend: Low: < 12"-•-Medium: 12–24"-•-Tall: > 24"

SPECIES by Common Name	COLOR								BLOOMING			HEIGHT		
	White	Pink	Red	Orange	Yellow	Blue	Purple	Foliage	Spring	Summer	Fall	Low	Medium	Tall
Mondo Grass	•					•	•			•		•		
Monkshood	•					•	•			•				•
Oriental Poppy		•	•	•					•	•			•	•
Ostrich Fern								•		•				•
Pachysandra	•								•			•		
Painted Fern/Lady Fern								•					•	•
Pasqueflower	•		•			•	•	•	•			•		
Penstemon	•	•	•			•	•		•	•	•	•	•	•
Peony	•	•	•		•		•		•	•				•
Phlox	•	•	•	•		•	•		•	•	•	•	•	•
Pinks	•	•	•				•		•	•		•	•	
Primrose	•	•	•	•	•	•	•		•	•		•	•	
Purple Coneflower	•	•		•	•					•	•		•	•
Sage	•	•				•	•		•	•	•		•	•
Sea Holly	•					•	•	•		•	•		•	•
Sedge						•	•			•			•	•
Sedum	•	•	•		•		•			•	•	•	•	
Shasta Daisy	•				•					•	•	•	•	
Solomon's Seal	•								•	•		•	•	•
Speedwell	•	•				•	•		•	•	•	•	•	•
Stokes' Aster	•	•			•	•	•			•	•			
Sunflower					•					•	•			•
Toad Lily	•	•					•			•	•			•
Trillium	•	•	•				•		•				•	
Tulip	•	•	•	•	•		•		•			•	•	•
Virginia Bluebells						•	•		•				•	
Wild Ginger						•	•		•	•		•		
Wood Fern								•					•	•
Yarrow	•		•		•		•	•		•	•	•	•	•

Sun	Part Shade	Light Shade	Shade	Moist	Well Drained	Dry	Fertile	Average	Poor	USDA Zones	Page Number	SPECIES by Common Name
•	•			•	•		•			6–10	252	Mondo Grass
•				•			•	•	•	3–8	254	Monkshood
•					•		•	•		3–7	258	Oriental Poppy
	•	•		•			•	•		1–8	262	Ostrich Fern
		•	•	•	•			•		4–9	264	Pachysandra
	•	•	•	•				•		4–8	266	Painted Fern/Painted Lady
•	•				•	•	•			3–7	268	Pasqueflower
•					•	•	•	•		4–8	270	Penstemon
•			•	•	•		•			2–7	274	Peony
•	•	•		•	•			•	•	3–8	280	Phlox
•					•	•		•	•	3–9	284	Pinks
	•	•		•	•			•		3–8	288	Primrose
•		•		•		•	•	•	•	3–8	292	Purple Coneflower
•					•			•		3–9	296	Sage
•					•		•	•		4–8	300	Sea Holly
•	•			•	•		•			6–9	302	Sedge
•					•	•		•		3–8	304	Sedum
•	•			•	•		•			4–9	308	Shasta Daisy
	•	•	•	•	•		•			4–8	310	Solomon's Seal
•				•	•			•		3–8	312	Speedwell
•					•	•	•	•		4–9	316	Stokes' Aster
•					•	•	•	•		4–8	318	Sunflower
	•	•	•	•	•		•			4–9	320	Toad Lily
	•	•	•	•	•			•		4–7	322	Trillium
•					•		•			3–8	324	Tulip
	•			•	•			•		3–7	328	Virginia Bluebells
	•		•	•	•	•		•		4–8	330	Wild Ginger
	•	•		•			•			3–8	332	Wood Fern
•					•	•		•	•	3–9	336	Yarrow

Glossary of Terms

acid soil: soil with a pH lower than 7.0

alkaline soil: soil with a pH higher than 7.0

annual: a plant that germinates, flowers, sets seed and dies in one growing season

basal foliage: leaves that form from the crown

basal rosette: a ring or rings of leaves growing from the crown of a plant at or near ground level; flowering stems of such plants grow separately from the crown

biennial: a plant that germinates and produces stems, roots and leaves in the first growing season; it flowers, sets seed and dies in the second growing season

crown: the part of a plant at or just below soil level where the shoots join the roots

cultivar: a cultivated plant variety with one or more distinct differences from the species, such as flower color, leaf variegation or disease resistance

damping off: fungal disease causing seedlings to rot at soil level and topple over

deadhead: to remove spent flowers to maintain a neat appearance and encourage a longer blooming period

direct sow: to plant seeds directly in the garden, where you want them to grow

disbud: to remove some flower buds to improve the size or quality of those remaining

dormancy: a period of plant inactivity, usually during winter or unfavorable climatic conditions

double flower: a flower with an unusually large number of petals, often caused by mutation of the stamens into petals

genus: a category of biological classification between the species and family levels; the first word in a Latin name indicates the genus

harden off: to gradually acclimatize plants that have been growing in a protective environment to a more harsh environment, e.g., plants started indoors being moved outdoors

hardy: capable of surviving unfavorable conditions, such as cold weather

humus: decomposed or decomposing organic material in the soil

hybrid: a plant resulting from natural or human-induced crossbreeding between varieties, species, or genera; the hybrid expresses features of each parent plant

knot garden: a formal design, often used for herb gardens, in which low, clipped hedges are arranged in elaborate, knot-like patterns

neutral soil: soil with a pH of 7.0

invasive: able to spread aggressively from the planting site and out-compete other plants

node: the area on a stem from which a leaf or new shoot grows

pH: a measure of acidity or alkalinity (the lower the pH, the higher the acidity); the pH of soil influences availability of nutrients for plants

perennial: a plant that takes three or more years to complete its life cycle; a herbaceous perennial normally dies back to the ground over winter

rhizome: a food-storing stem that grows horizontally at or just below soil level, from which new shoots may emerge

rootball: the root mass and surrounding soil of a container-grown plant or a plant dug out of the ground

runner: a modified stem that grows on the soil surface; roots and new shoots are produced at nodes along its length

semi-double flower: a flower with petals that form two or three rings

side-dressing: applying fertilizer to the soil beside or around a plant during the growing season to stimulate growth

single flower: a flower with a single ring of typically four or five petals

species: the original species from which cultivars and varieties are derived; the fundamental unit of biological classification

subspecies (subsp.): a naturally occurring, regional form of a species, often isolated from other subspecies but still potentially interfertile with them

taproot: a root system consisting of one main root with smaller roots branching from it

tender: incapable of surviving the climatic conditions of a given region and requiring protection from frost or cold

true: the passing of desirable characteristics from the parent plant to seed-grown offspring; also called breeding true to type

tuber: the thick section of a rhizome bearing nodes and buds

variegation: foliage that has more than one color, often patched or striped or bearing differently colored leaf margins

variety (var.): a naturally occurring variant of a species; below the level of subspecies in biological classification

Index of Plant Names

Names in **bold** are main plant entries.